iPhone 17 Pro User Guide

A Simplified Manual with Hidden Features, Tips &

Tricks, Camera Settings, and iOS 18 Essentials for

Seniors & Beginners

Jamie Bernard

Disclaimer

This book is an independent publication and is not authorized, sponsored, or endorsed by Apple Inc. The word "iPhone" and related product names are trademarks of Apple Inc., registered in the United States and other countries. All other product names and brands mentioned are the property of their respective owners.

The information contained in this guide is provided for educational and reference purposes only. While every effort has been made to ensure accuracy at the time of writing, technology evolves rapidly, and updates or changes to Apple's software and hardware may render some content out of date. The author and publisher make no representations or warranties regarding the completeness, reliability, or accuracy of the information in this book.

By using this guide, you acknowledge and agree that the author and publisher shall not be held liable for any direct, indirect, incidental, or consequential damages arising from the use of this material. Always follow Apple's official documentation and consult Apple

Support or an authorized service provider for technical issues that cannot be resolved safely at home.

First Edition, 2025

Why This Book Exist

Every September, the world gathers around the glow of a screen to watch Apple unveil another masterpiece—the newest iPhone. There's always that first rush of excitement: faster chip, better camera, sleeker design, smarter features. But after the applause fades, something very human happens. You open the box, feel that smooth glass for the first time, power it on… and within minutes, you're lost in an ocean of icons, toggles, and notifications you don't fully understand.

You're not alone. Every year, millions of iPhone users—new and seasoned—find themselves more confused, not less. And there's a reason for that.

The Truth About Why New iPhone Users Struggle More

Each Year

Technology isn't getting simpler—it's getting *smarter*. And while Apple is famous for its clean design and intuitive experience, that simplicity often hides incredible complexity beneath the surface.

Each new iPhone release isn't just an upgrade—it's a reimagining of how your phone thinks, learns, and interacts with the world. With every new iOS version, Apple adds layers of automation, personalization, and hidden features meant to make life easier. But for most users, these layers pile up like unseen bricks, turning what should feel "smart" into something overwhelming.

A few years ago, you could open your iPhone, set your ringtone, and feel done. Now, you have:

- *Multiple* ways to back up and sync (iCloud, Finder, AirDrop, Quick Start, and more)
- Dozens of background processes draining battery or using data you never knew existed

- Privacy pop-ups that appear out of nowhere asking you to "Allow While Using App"

- Motion effects that make icons shimmer but leave some users dizzy or tired

- Photo formats (HEIC, ProRAW, ProRes) that even professionals debate about

- Notifications layered in "summaries," "focus modes," and "scheduled delivery"

And that's not even counting when things go wrong—like battery drain, overheating, or apps crashing after an iOS update.

The truth is, Apple builds technology for everyone—but most people still need a human bridge between "smart" and *simple*. That's where this book comes in.

What This Guide Will Help You Achieve

This guide was written to restore what technology quietly took away: *confidence, ease, and control.*

7

It's not another generic manual filled with tech jargon or copied screenshots. It's a *human* companion—a book that talks to you like a friend who's already figured it out and wants you to feel capable.

By the time you finish, you will:

1. **Feel Confident, Not Clueless.**

   You'll know what every key setting does—and why it matters. You'll stop guessing and start understanding. When something goes wrong, you won't panic; you'll know where to look and what to try.

2. **Use Your iPhone with Ease.**

   No more fear of pressing the wrong button. You'll glide through apps, settings, and updates effortlessly. Whether you're taking photos, connecting to Wi-Fi, or organizing your files, you'll do it naturally, without second-guessing yourself.

3. **Stay in Control.**

   Your iPhone will stop feeling like it's in charge of you.

You'll learn how to take back control of notifications, privacy, battery life, and performance—so your phone serves your life, not the other way around.

You'll also discover hidden gems Apple never advertises—simple shortcuts, power-saving habits, and small settings that make your device more personal, efficient, and fun to use.

This book isn't about *keeping up with technology*; it's about *making technology keep up with you.*

How to Use the Visual, Step-by-Step Layout for Fast Results

Every page in this guide is built for *clarity and speed.*

You won't find long technical lectures or confusing tech slang. Instead, you'll find a *visual learning flow*—carefully structured so you can flip to any section and fix a problem in minutes.

Here's how to make the most of it:

1. **Follow the Icons.**

   Each section includes easy-to-spot visual cues:

   - *Setup Steps* — for first-time tasks or essential configurations

   - *Fix It Now* — fast troubleshooting you can do immediately

   - *Pro Tip* — advanced or hidden features most users miss

   - *Battery Saver* — practical ways to extend power

   - *Avoid This* — common mistakes or myths to skip

2. **Look, Read, Do.**

   Every instruction is paired with clear screenshots or visual markers—so you can match what you see on your screen to what's in the book. You don't need to memorize anything; you just act along.

3. **Jump Around—No Linear Reading Required.**

   This isn't a novel. You can open it anywhere—Battery,

Camera, Wi-Fi—and start. Each chapter stands on its own, yet together they form a complete mastery map.

4. **Learn by Solving.**

Every chapter includes real-world scenarios ("Why is my phone hot?" "Why are my photos blurry?" "Why does Wi-Fi keep disconnecting?") followed by exact fixes. You'll *learn while you repair*, which makes knowledge stick effortlessly.

5. **Use the Quick-Reference Index.**

At the back of the book, you'll find flowcharts and one-page summaries that condense entire topics—so if your phone misbehaves, you can turn there and act in seconds.

This isn't a manual you read once and shelve—it's a lifelong companion you'll return to whenever something feels off.

In Essence

This book exists because technology should *serve* you, not intimidate you. Because owning a powerful iPhone should feel empowering, not exhausting. And because every user—whether a beginner, a senior, a creator, or a business owner—deserves a clear, friendly path to mastery.

If you've ever felt like your phone was "too advanced," this guide will remind you that *you're still smarter than your smartphone.*

Table of Contents

Preface

Master your iPhone 17 Pro with the most complete, easy-to-follow, and visually rich guide ever written—crafted especially for seniors, beginners, and non-tech-savvy adults. This iPhone 17 Pro user guide is your trusted companion to understanding every feature, shortcut, and hidden gem inside Apple's smartest device yet. Whether you're setting up your phone for the first time or looking to uncover iPhone 17 hidden features, this all-in-one Apple iPhone 17 Pro handbook walks you through everything with crystal-clear steps and full-color illustrations.

Inside this iPhone 17 Pro beginners guide, you'll find a complete step-by-step walkthrough that makes the learning process simple, fun, and frustration-free. From the moment you unbox your device, this iPhone 17 Pro setup guide shows you how to get connected, personalize settings, protect your privacy, and explore all the amazing tools built into iOS 18. Every page is packed with iPhone

17 Pro tips and tricks, making it the ultimate iPhone 17 Pro how-to book for everyday users.

Discover everything Apple didn't tell you — including iPhone 17 Pro hidden settings, advanced iPhone 17 Pro privacy settings, Face ID setup, Wi-Fi and Bluetooth setup, and Control Center explained. You'll also master gestures, learn the smartest iPhone 17 Pro shortcuts, and gain access to battery saving tricks, camera and photo settings, and accessibility features that make this iPhone 17 simplified manual perfect for everyone — especially those who prefer large print seniors editions and step-by-step visuals.

Learn how to fix everyday problems confidently with a full iPhone 17 Pro troubleshooting and Apple iPhone troubleshooting section that includes how to solve iPhone 17 battery problems, fix iPhone 17 Pro overheating, perform a not charging fix, stop app crashing, and reset iPhone 17 safely. With built-in guidance for iPhone 17 Pro backup and restore, data recovery, and how to transfer data to a new iPhone 17, you'll always be in control of your digital life.

Photography lovers will appreciate the iPhone 17 Pro camera guide and iPhone 17 Pro photography tips, which reveal how to take better photos, adjust photo settings, and explore the camera's professional-grade features. You'll also find iPhone 17 Pro maintenance tips, iPhone 17 Pro iOS 18 update guide, and iPhone 17 Pro security guide to keep your device optimized for years to come.

This iPhone 17 Pro user manual step-by-step is written in plain English with an easy flow for learners of all ages — a truly easy iPhone 17 guide for seniors, a complete beginner's guide to iPhone 17, and a simplified iPhone 17 book for elders. Whether you're following an iPhone 17 tutorial for older adults, reading through iPhone 17 step-by-step instructions, or exploring this iPhone 17 Pro illustrated guide, you'll gain the confidence to use your iPhone 17 Pro effectively every single day.

Perfect for users everywhere — this iPhone 17 Pro guide UK edition, manual Canada, training USA edition, tips Australia, and setup Africa / Nigeria adaptation ensures worldwide usability. It's more than just an iPhone 17 Pro Apple help book — it's an iPhone 17 Pro complete guide for seniors and a global manual for elders and beginners, including those looking for an iPhone 17 for grandparents made easy edition.

With bonus sections on voice and Siri commands, Siri guide, and iPhone 17 Pro basics explained, this iPhone 17 Pro easy instructions resource also teaches you how to speed up iPhone 17 Pro, perform iPhone 17 battery fix, handle Wi-Fi issues, and manage app crashing guides like a pro. It's the most comprehensive iPhone 17 Pro for older adults and accessibility tips manual available today — a complete iPhone 17 Pro how-to guide and beginners handbook designed to help you learn iPhone 17 step by step and master your iPhone 17 Pro easily.

This iPhone 17 made simple, iPhone 17 for seniors large print, iPhone 17 for seniors simplified, and iPhone 17 for seniors worldwide edition is tailored for those who want real understanding — not technical jargon. It's your iPhone 17 Pro troubleshooting and repair book, your personal iPhone 17 Pro hidden secrets explained, and your go-to iPhone 17 Pro simplified manual for seniors.

Whether you're in London, Lagos, Toronto, or Los Angeles, this Apple iPhone 17 Pro handbook is your lifetime learning companion — your iPhone 17 Pro step-by-step beginners handbook, and your all-in-one iPhone 17 Pro complete walkthrough. Simple, powerful, and filled with real solutions, this guide will help you transform confusion into confidence and make your iPhone 17 Pro feel truly yours.

Introduction

Welcome to the iPhone 17 Pro Era

Every generation of the iPhone has been a conversation between innovation and expectation—a dialogue between what technology can do and what human beings actually need. With the arrival of the *iPhone 17 Pro,* that conversation reaches a new intensity. This is not just another smartphone. It's a pocket-sized supercomputer with enough intelligence, processing power, and photographic brilliance to rival professional tools once reserved for studios and labs. But with great power comes great complexity—and that's precisely where most users begin to feel left behind.

Let's take a closer look at what this new era brings, why even the tech-savvy sometimes get lost in it, and how this book was designed to help you not just *own* the iPhone 17 Pro—but *master* it.

What's New in iPhone 17 Pro

The iPhone 17 Pro represents the most refined expression of Apple's technological philosophy: sleek, silent power wrapped in precision engineering. Yet beneath that familiar glass-and-titanium frame lies a set of upgrades that quietly redefine what's possible in a handheld device.

1. The A17 Bionic Evolution

(Apple's naming convention continues from the A16, but we'll refer to it here as the *A17 Bionic*—the next step in the lineage.)

Built on an advanced 3-nanometer architecture, the chip delivers lightning-fast responsiveness, dramatically improved power efficiency, and up to 30% better GPU performance. It's not just faster—it's *smarter*. The A17 integrates enhanced neural engines that enable real-time learning and adaptive system management. In simple terms, your iPhone now anticipates your habits: it knows which apps you'll open, optimizes background processes before you even tap, and adjusts performance to match your daily rhythm.

2. Advanced Cooling System

Apple has finally addressed one of the long-standing concerns: heat. The iPhone 17 Pro introduces a re-engineered vapor-chamber cooling design that dissipates thermal buildup during high-intensity use—whether that's recording 4K video, gaming, or multitasking with power-hungry apps. For creators, this means longer recording sessions without performance throttling. For everyday users, it means fewer overheating warnings, longer battery life, and a cooler, more comfortable experience in hand.

3. The Camera Sensor Revolution

Photography has always been Apple's crown jewel, and the iPhone 17 Pro raises the bar yet again. The new *1-inch stacked CMOS sensor* captures more light than ever before, delivering richer colors, deeper shadows, and clearer details even in low-light environments. The upgraded computational engine allows *instant HDR fusion*—merging exposures in real time rather than after capture—giving photos a level of realism that feels cinematic. Add to that the ability to shoot 8K ProRes video and advanced lens coatings that minimize

24

glare, and you're holding a device that challenges DSLRs in the palm of your hand.

4. iOS 18's Hidden Shifts

At first glance, iOS 18 looks like a gentle evolution of iOS 17—but under the surface, Apple has restructured the experience around *context awareness* and *adaptive automation*. Here's what's subtly changed:

- **Smarter Control Center:** now fully customizable, allowing widgets, toggles, and app shortcuts arranged your way.

- **Enhanced Focus Modes:** smarter scheduling and AI-driven suggestions that silence distractions before they reach you.

- **"Liquid Glass" Design:** dynamic transparency and motion effects that give depth—but also strain some users' eyes (we'll cover how to adjust this later).

- **Expanded Privacy Dashboard:** easier visibility of what's tracking you and when, with instant one-tap permission toggles.

- **Battery Intelligence:** iOS now predicts power demand and adjusts background refresh based on time, location, and app patterns.

These features combine to make your iPhone feel more personal and alive—but they also introduce new layers of settings, behaviors, and subtleties that can easily confuse even experienced users.

Why Even Tech-Savvy Users Get Overwhelmed

If you've used iPhones for years, you might assume that each upgrade would feel familiar. But Apple's philosophy has evolved from *manual control* to *machine-assisted intelligence*. The phone that once waited for your command now acts almost autonomously—preloading apps, filtering notifications, adjusting exposure, saving battery, even determining when it thinks you want silence.

For the average user, that's both impressive and intimidating.

Here's the paradox: *simplicity now hides complexity.*

The same clean interface that makes the iPhone appear "easy to use" is layered with background processes, automated settings, and invisible decisions that affect everything from performance to privacy. Tap the wrong switch, and you may disable a feature you didn't even know existed. Change one Focus Mode, and your calls stop coming through. Enable a photo format, and your images no longer open on your computer.

Even the most tech-comfortable users find themselves asking:

- "Why is my battery draining overnight?"
- "Why does my phone feel hotter than before?"
- "Why did my photos suddenly disappear from the gallery?"
- "Why can't I AirDrop to my Mac anymore?"

The truth is, Apple's design goal is to anticipate problems before you notice them—but that same automation can make it harder to understand what's actually happening behind the scenes. Without the right guidance, it's easy to feel like the phone is running *you* instead of the other way around.

That's why you need a bridge—a practical, human-centered guide that turns hidden complexity into simple clarity.

How This Book Differs from Apple's Official Guide

Apple's official manuals are precise, elegant, and free—but they're written for the system, not for *you*. They describe features, not experiences. They explain *what* exists but rarely tell you *why it matters* or *what to do when it doesn't work.*

This book takes the opposite approach. It's a real-world field guide built from actual user behavior, tested solutions, and day-to-day scenarios—not corporate documentation. Here's how it's different:

1. **Real Use, Not Theoretical Features.**

   Every chapter is grounded in how people *actually* use their phones—at home, at work, on the road, under poor network conditions, or in countries where Apple's services behave differently. You'll learn what happens when your Wi-Fi drops mid-FaceTime, or how to stop your camera overheating on a long recording, or why your iPhone won't charge properly in a hot car.

2. **Real Language, Not Tech Jargon.**

You won't find dense terms like "neural engine task offloading" or "thermal throttling subroutine." Instead, you'll find clear, human explanations: what it means, how it affects you, and what to do about it. The tone is conversational—like a friend who's good with tech, sitting beside you and guiding you through it.

3. **Real Fixes, Not Just Features.**

Apple's guide tells you how to enable something. This book tells you what to do when that feature misbehaves—or when it slows your phone, drains your battery, or disrupts your day. Each section is built around the *Fix–Understand–Optimize* cycle:

o **Fix:** Immediate, proven steps to solve a problem.

o **Understand:** A short explanation of why it happened.

29

o **Optimize:** Proactive tips to prevent it from happening again.

4. **Global Perspective.**

Many tech manuals assume the reader lives in regions with perfect networks and consistent Apple support. This guide considers realities—like inconsistent data coverage, limited repair centers, or regional pricing for iCloud and AppleCare. Whether you're in Aberdeen, Lagos, Toronto, or Sydney, you'll find context that speaks your language.

5. **Visual Clarity and Speed.**

Every fix, setting, or process is explained through clear visuals and step-by-step directions you can follow at a glance. You don't need to read an entire chapter to solve a problem—you can flip to the relevant page, act immediately, and get back to your day.

In Simple Terms

The *iPhone 17 Pro Era* isn't just about faster chips or better cameras—it's about living with a device that learns from you, predicts your needs, and quietly manages more of your digital life than ever before. That's both extraordinary and challenging.

This book was written to help you navigate that frontier with calm, confidence, and control. To remind you that behind every algorithm is still a human decision—and that the smartest technology is the one that bends to your will, not the other way around.

So, welcome to the iPhone 17 Pro era. Let's make it yours.

Part I – Getting Started the Right Way

(Prevent 80% of user problems before they start.)

Chapter 1

The Smart Setup Blueprint

The first hour you spend with your new iPhone 17 Pro will shape how smoothly it runs for years to come. Think of setup not as a quick formality, but as the foundation of your entire digital life. When done right, it saves you from battery drain, random crashes, endless notifications, and the frustration of constantly tweaking settings later.

This chapter walks you through that process deliberately—step by step, decision by decision—so that your iPhone begins its life *optimized*, not overburdened.

Setting Up Your iPhone 17 Pro Without Headaches

Unboxing an iPhone should feel exciting, not confusing. Yet many users rush through setup screens, tapping "Continue" without realizing they've just enabled features they don't need—features that quietly drain battery, eat storage, and collect data they never use.

The key to a stress-free setup is *pace and purpose*. Every choice you make in these early minutes is a seed; plant them wisely, and your iPhone will reward you with longevity and reliability.

Step 1: Power On and Prepare Your Environment

- Plug your phone into power during setup. Initial updates and app restores consume more battery than normal.

- Connect to a strong, private Wi-Fi network—not public hotspots—to avoid interruptions.

- Keep your old device, Apple ID, and passwords nearby. You'll need them for authentication and data migration.

Step 2: Update Before You Migrate

Before transferring data, check for the latest version of *iOS 18.* Apple often releases a small post-launch patch that fixes setup bugs.

Go to *Settings › General › Software Update* and install any update available. This single step prevents hours of troubleshooting later.

Choosing the Right Restore Method

Apple offers three main ways to bring your old life onto your new phone: *Quick Start, iCloud Restore,* and *Cable Transfer*. Each has strengths, weaknesses, and ideal use cases.

1. Quick Start — Fast, Wireless, and Seamless

When your old iPhone is nearby, Quick Start uses Bluetooth and Wi-Fi Direct to clone your apps, settings, and layout.

- **Pros:** Convenient, nearly complete transfer (including wallpaper, home screen arrangement, and Apple Pay setup).
- **Cons:** Can stall if your Wi-Fi drops; sometimes brings along old cached files that carry over minor bugs.

Best for: Newer iPhones on iOS 16 or later, with strong Wi-Fi and recent backups.

2. iCloud Restore — Best for Clean Starts

If your old phone is already backed up to iCloud, you can restore directly from that cloud image.

- **Pros:** Doesn't require both phones to be side by side. Keeps essential data without unnecessary clutter.

- **Cons:** Dependent on network speed; apps re-download individually, which can take hours.

Best for: Users with limited local storage, or when your old phone is unavailable or broken.

3. Cable Transfer — The Purist's Option

Using a Lightning-to-USB-C or USB-C-to-USB-C cable (depending on your old model) allows a direct, lossless, offline transfer.

- **Pros:** Fastest and most complete; avoids Wi-Fi interruptions.

- **Cons:** Requires both devices and a compatible cable; slightly more technical to initiate.

Best for: Power users, photographers, and professionals transferring large photo libraries.

Pro Tip: If your old phone has been acting buggy—crashing apps, battery drain, unexplained heat—avoid cloning those issues. Back up *only* data (via iCloud or Finder) and set up the iPhone 17 Pro as *new*. Then re-download your essential apps manually. It takes longer, but you start fresh, free from inherited glitches.

Hidden iCloud Settings That Save Battery

Most users assume iCloud is "set and forget." But behind that seamless sync are background processes constantly uploading, downloading, and indexing your files. Left unchecked, they can quietly eat your data and battery.

Here's how to tune iCloud for balance—efficiency without compromise.

1. Disable Unnecessary iCloud Backups

Go to *Settings › [Your Name] › iCloud › iCloud Backup.*

- Toggle off backups for devices you no longer use.

- Schedule backups while charging overnight.

2. Manage App-Specific Syncing

Navigate to *iCloud › Apps Using iCloud.*

Deselect heavy apps like GarageBand or large document editors unless you rely on cross-device syncing.

3. Optimize Photos the Smart Way

Enable *Optimize iPhone Storage* under Photos.

This stores lightweight versions on your device and full-resolution copies in the cloud—preserving quality without filling storage.

4. Turn Off Private Relay (if not essential)

Apple's Private Relay is great for privacy but slightly increases background traffic. If battery is your top priority, disable it under *Settings › iCloud › Private Relay.*

5. Use "Data Saver" Mode

Under *Settings › Cellular › Cellular Data Options, enable Low Data Mode.* It pauses background syncing until Wi-Fi is available—ideal in areas with weak coverage or high data costs.

Quick Save:

After finishing setup, reboot your iPhone. This resets iCloud's background indexing and prevents overnight battery spikes during the first 24 hours.

Wi-Fi, SIM, and eSIM Setup Done Right

Connectivity issues often start with rushed setup. Follow these best practices:

1. Wi-Fi Precision

- Always connect to a *secured 2.4 GHz or 5 GHz network*, never public hotspots during activation.

- Disable *Auto-Join Hotspots under Settings › Wi-Fi › Auto-Join Hotspot* to prevent random reconnections that drain battery.

- Tap the (i) next to your Wi-Fi name › *Private Wi-Fi Address* → *ON* to improve privacy.

2. SIM vs eSIM: Choose Intelligently

The iPhone 17 Pro supports *dual eSIMs* or one physical SIM + one eSIM.

- **Physical SIM:** Simple, easily swappable between devices; best if your carrier doesn't support eSIM.

- **eSIM:** Faster activation, supports dual lines, eliminates card wear. Perfect for international travel—just add a data plan digitally.

Set Up eSIM:

Go to *Settings › Cellular › Add eSIM,* scan the QR code from your carrier, and wait for activation. Reboot once connected.

3. Network Stability Tip

Immediately after activation, go to *Settings › General › About.*

If prompted, install *Carrier Settings Update.* This small update ensures the best performance for calls, data, and messaging.

The Essential First-Hour Checklist

(Avoid the traps that cause later crashes, heat, and lag.)

Your first hour determines whether your iPhone runs like silk or fights you at every update. Complete this checklist before diving into customization:

Battery Health Setup

- Enable *Optimized Battery Charging* under *Settings › Battery › Battery Health & Charging.*
- Avoid restoring while charging overnight on Day 1—setup processes generate heat. Charge after migration is complete.

Connectivity & Updates

- Update carrier settings (as above).

- Install pending *Software Updates* before restoring apps.

- Turn off *Automatic Downloads* temporarily under *App Store › Automatic Downloads*—re-enable later when stable.

iCloud Control

- Confirm backup status (Settings › Your Name › iCloud › iCloud Backup › Back Up Now).

- Disable unnecessary app syncing (as explained earlier).

- Check *Find My iPhone › ON* and ensure location access is granted.

System Optimization

- Clear temporary setup cache: perform a *soft reboot* (Volume Up → Volume Down → Hold Side Button).

- Go to *Settings › Privacy › Analytics & Improvements*, toggle off "Share iPhone Analytics" to reduce background reporting.

- Turn off "Significant Locations" under *Privacy › Location Services › System Services* to cut background GPS drain.

Camera & Storage Basics

- Open the Camera app once—it initializes photo caches properly.

- In *Settings › Camera › Formats*, choose *High Efficiency* for balanced quality and space.

- Enable *iCloud Photos* if you intend to use it; disable if you prefer manual control.

Notifications & Focus Modes

- Before notifications flood in, set a *Default Focus Mode* under *Settings › Focus.*

- Disable unnecessary app notifications—doing this early prevents future overwhelm.

Memory Management

- Don't restore too many heavy apps simultaneously. Let the first wave install, then add the rest later. This avoids overheating and failed downloads.

A Thought Before You Move On

Your iPhone 17 Pro is a marvel of engineering—thousands of micro-decisions happening every second to make your life smoother. But those decisions begin with the ones *you* make today. Setting up wisely doesn't just prevent problems—it teaches your phone how to serve you efficiently from day one.

After completing this Smart Setup Blueprint, your device will already be running cooler, cleaner, and faster than most. The groundwork is laid; from here on, every tap will feel effortless.

Chapter 1: The Smart Setup Blueprint

Setting Up Your iPhone 17 Pro Without Headaches

1 Stetup step 1

- Plug it into power during setup Connect to a strong, private Wi-Fi network, and keep cld
- Old device, Apple iD, and passwords nearby

↑ **Update: iQS 18**

Before migrating

Choosing the Right Restore Method (Quick Start vs iCloud vs Cable)

- Quick Start
  Fast, wireless, and naay complete
- Install pending iOS 18

Choosing the Right Restore Method
(Quick Start vs iCloud vs Cable)

- Quick Start - best for clean starts
- Optimaze on network speed
- Cable Transfer
  Fast, seat Privaty's option
- Turn off Private Relay if essential

Hidden iCloud Settings That Save Battery

- Disable unnecessary iCloud backups
  Settings > (Your Name) > iCloud
  iCloud Backup
- Manage app-speecific syncing
  Settings > Photos > Optimus peed
- Optimize Photos the smart way
  Settings > Photos > Optimize
  iPhone Storage, Stores
  lighweight versions on b your device
  and full-resolution copies
- Turn off Private Relay if not essental

The Essential First-Hour Checklist — Avoid the traps that cause later crashes and overheating

- Battery Health setup + enable Optimized Battery Charging
- Install pending Software Updates before restoring apps
- Disable unnecessary app sycing (confirm backup status
- Turn off "Significant Locations" for GPS drain
- Soft reboct; Volume Up ＞ Volume Down – Hold Side button
  ○ Soff reboot, Soft, reboot
  Volume Up – Volume Down
  Hold Side button

11

45

Chapter 2

Mastering the New Interface in iOS 18

Every few years, Apple quietly reshapes the way we interact with the iPhone—never with a loud revolution, but with small, deliberate evolutions that redefine the entire experience. iOS 18 is one of those quiet revolutions. On the surface, it feels familiar. But as you begin to swipe, tap, and scroll, you'll realize the rhythm has changed.

The gestures feel smoother. The visuals have a new sense of depth. Notifications appear softer and less intrusive. And under it all, the system has become more intuitive, more adaptive—yet also, for many, more overwhelming.

This chapter is your map through that new terrain. You'll learn how to control what's new, simplify what feels too busy, and use the latest tools with comfort and clarity.

Understanding the Redesigned Control Center,

Notification Summary, and App Library

1. The All-New Control Center

Control Center is no longer just a pull-down menu—it's your command deck. Apple redesigned it to be fully customizable, layered, and contextual.

Access: Swipe down from the top-right corner of your screen.

What's New:

- **Multiple Pages:** You can now swipe left or right within Control Center to reveal additional panels (Media, Home, Connectivity, Battery, etc.).

- **Custom Layouts:** Long-press anywhere in Control Center and tap *Edit Controls.* Drag icons to rearrange or delete those you don't need.

- **New Widgets:** Add mini versions of essential apps— Flashlight, Focus Modes, Calculator, Notes, Screen

Recording, Low Power Mode, and even third-party app controls.

- **Adaptive Brightness Slider:** Now adjusts both brightness *and tone warmth* in real time based on ambient light, helping reduce glare during long screen time.

Pro Tip:

Hold any icon for two seconds to open *expanded controls.* For instance, long-press the Wi-Fi icon to switch networks directly, or long-press Bluetooth to pair new devices without diving into Settings.

2. Notification Summary—Now Smarter, Softer, and Saner

Notifications used to be noisy—constant pings that fractured your focus. In iOS 18, Apple refined the Notification Summary to help you regain peace.

Here's what's new:

- **Smart Grouping:** Notifications from the same app or category (social, work, deliveries) are bundled intelligently, so your Lock Screen doesn't look like a battlefield.

- **Adaptive Priority:** iOS 18 now learns which apps you actually interact with and delays less-important alerts automatically.

- **Morning & Evening Digests:** You can now schedule summaries twice a day—say, at 8 AM and 8 PM—to review everything calmly, instead of getting constant interruptions.

To Set It Up:

Go to *Settings › Notifications › Scheduled Summary.*

Enable it and choose which apps belong in your digest. Then schedule preferred delivery times.

Pro Tip:

Use *Time-Sensitive Notifications* for only your top three essential apps (e.g., Messages, Calendar, Health). This keeps critical alerts coming through even in Focus Mode.

3. App Library — Simpler, But More Intelligent

The App Library—found by swiping all the way left past your last Home Screen—has matured.

It now uses *predictive grouping* powered by machine learning. Apps you use together (e.g., Camera + Photos + Instagram) appear side by side automatically.

New in iOS 18:

- **Pinned Categories:** Long-press any app folder and select *Pin to Top.* Perfect for fast access to "Utilities" or "Work."

- **Search That Thinks Like You:** Instead of typing full app names, you can type what the app *does.* For instance, type "video edit," and it will surface CapCut or iMovie.

- **App Hiding 2.0:** Apps hidden from the Home Screen now have a discrete space in the App Library under *Hidden Apps.* Access requires Face ID.

Quick Shortcut:

Swipe down once in App Library, and a small alphabetical grid appears on the right—perfect for fast scrolling by first letter.

"Liquid Glass" and Motion Effects — How to Reduce Eye Strain and Vertigo

One of iOS 18's most beautiful and controversial updates is its *Liquid Glass* design—a dynamic interface that subtly moves, bends, and reflects light as you scroll.

It looks stunning in promo videos, but for some users, especially those sensitive to motion or bright contrasts, it can cause mild disorientation, eye strain, or even vertigo.

If that's you—or you simply prefer a calmer screen—here's how to balance aesthetics and comfort.

1. Reduce Motion Effects

Go to *Settings › Accessibility › Motion › Reduce Motion* and toggle it on.

This limits parallax (the floating icon effect), smooths transitions, and stops wallpapers from shifting as you tilt your device.

Bonus: It also saves battery, since the GPU performs fewer animations.

2. Disable Liquid Glass Transparency

Navigate to *Settings › Accessibility › Display & Text Size › Reduce Transparency.*

When turned on, backgrounds in Control Center and widgets lose the "frosted glass" blur, becoming easier to read in bright light.

3. Adjust White Point and Color Tint

Under *Display & Text Size,* scroll to *Reduce White Point* and lower it by 20–30%.

This dims overly bright elements without reducing screen brightness overall—a huge help for eye comfort during nighttime use.

You can also activate *Color Filters › Tint › Soft Warm* to soften the cold blue tones that often cause fatigue.

4. Use True Tone and Night Shift Wisely

In *Settings › Display & Brightness*, enable *True Tone* (adapts screen color to your environment) and *Night Shift* (reduces blue light after sunset).

Adjust Night Shift's schedule from "Sunset to Sunrise" or custom hours that match your bedtime routine.

5. Wallpaper Choices Matter

Avoid high-contrast or animated wallpapers if prone to motion sickness. Choose muted gradients or still backgrounds instead. A calm wallpaper keeps icons steady, reducing micro-movements that can trigger discomfort.

Accessibility Features Everyone Should Use

Apple's accessibility tools aren't just for those with special needs—they're *performance enhancers* for everyone. They simplify, clarify, and personalize your experience so your phone adapts to *you*.

1. Reduce Motion (again)

As covered earlier, it's both an accessibility and wellness feature. Beyond preventing dizziness, it also makes transitions faster—great for anyone who prefers efficiency over flair.

2. Display Accommodations

Found under *Settings › Accessibility › Display & Text Size.*
Here's what each tool can do for you:

- **Increase Contrast:** Deepens text and outline visibility in bright environments.
- **Bold Text:** Instantly improves readability for messages and menus.

- **Smart Invert:** Flips interface colors for a dark-mode experience even in apps that don't support it.

- **Color Filters:** Essential for color-blind users, but also great for reducing harsh blues or whites that cause glare.

Pro Tip:

Try setting *Reduce White Point = 25% and Bold Text = ON* for a comfortable everyday balance between clarity and eye comfort.

3. Voice Control and Dictation

iOS 18 refines *Voice Control* so it now recognizes contextual commands more naturally. You can say "Scroll down a bit," "Open Camera," or "Take a screenshot," and it executes instantly.

For text input, dictation now runs entirely on-device (faster, private, no internet required).

4. AssistiveTouch

Ideal for those who dislike pressing physical buttons.

Go to *Settings › Accessibility › Touch › AssistiveTouch* and toggle it on. A floating button appears, giving one-tap access to volume, multitasking, and screenshots.

You can customize it to hold shortcuts like "Lock Screen" or "Restart."

Navigation Secrets for Beginners: Swipes, Gestures, and Shortcuts Apple Hides in Small Print

Apple's gestures are elegant—but many are never explained anywhere obvious. Here's a guide to the ones you'll actually use daily.

1. Home & Multitasking

- **Return Home:** Swipe up from the bottom edge and pause to go home.

- **App Switcher:** Swipe up halfway and pause—then slide sideways to switch apps.

- **Quick App Swap:** Swipe *left or right* along the bottom edge to toggle between your most recent apps instantly.

2. Control Center & Notifications

- **Control Center:** Swipe down from the top-right corner.

- **Notification Center:** Swipe down from the top-left corner or middle of the top edge.

- **Search/Spotlight:** Swipe down on the Home Screen itself.

3. Screenshot & Quick Notes

- Press *Side Button* + *Volume Up* simultaneously for a screenshot.

- After capture, tap the thumbnail → choose *Add to Quick Note.*

- You can also swipe diagonally from the bottom-right corner (using Apple Pencil or finger) to create a Quick Note.

4. Drag & Drop Magic

Tap and hold an image, link, or block of text until it "lifts," then use another finger to swipe into a different app—like Notes or Messages—and drop it. iOS 18 handles this multi-touch seamlessly.

5. Hidden Keyboard Shortcuts

- **Double-Tap Space Bar:** Inserts a period automatically.

- **Hold Space Bar:** Turns the keyboard into a trackpad for precise cursor movement.

- **Swipe Down on Keys:** Access numbers and symbols instantly (on smaller keyboards).

6. Swipe Gestures in Safari & Photos

- **Swipe Right:** Go back.

- **Swipe Left:** Go forward.

- **Pinch Out (Photos):** Zoom into details; *Pinch In:* Quickly return to gallery view.

7. Quick Actions on Home Screen

Long-press any app icon for contextual menus—Camera lets you jump straight to "Selfie" or "Portrait Mode," Messages offers "New Message."

You can customize these in *Settings › Siri & Search › Suggested Shortcuts.*

Bringing It All Together

Mastering iOS 18 isn't about memorizing every new feature—it's about *personalizing the experience so it feels invisible.* The redesigned Control Center gives you authority over your daily actions. The smarter Notification Summary restores peace of mind. The Liquid Glass aesthetic, when tuned correctly, can feel luxurious rather than overwhelming. And the accessibility features are your silent allies, making everything clearer, calmer, and kinder to your eyes.

Once you've set your preferences, gestures, and comfort levels, your iPhone 17 Pro stops being a machine and starts becoming a natural extension of your hand.

Long press
to expand

NOTIFICATION

Today

Nessages 20 >
2 alerts

News 2h >
2 alerts

2 MORE NOTIFICATIONS

Reduce Transparency

Reduce Transparency

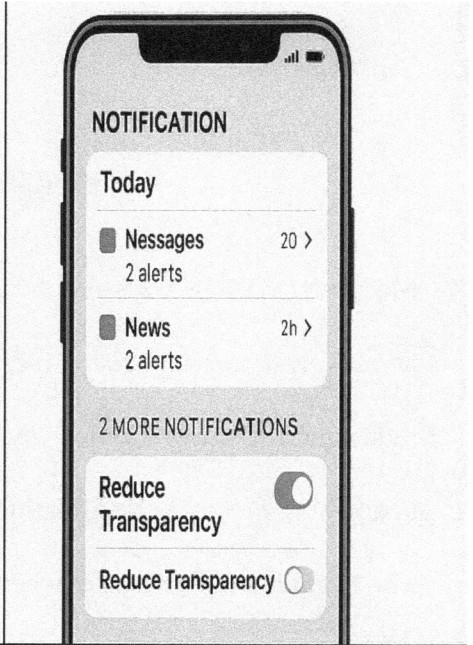

"LIQUID GLASS" AND MOTION EFFECTS

< Accessibility

MOTION

Reduce Motion

Auto-Play Message Effects

Auto-Play Video Previews

REDUCE TRANSPARENCY

Display Accommons

NAVIGATION SECRETS

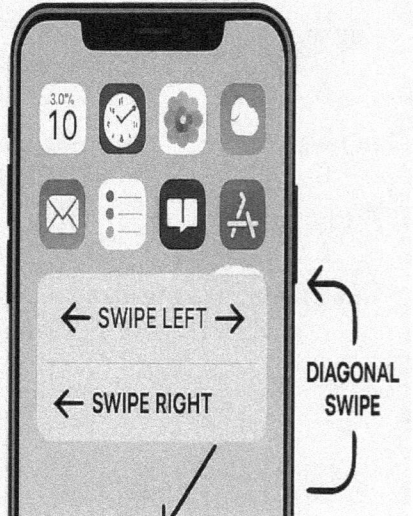

3.0%
10

← SWIPE LEFT →

← SWIPE RIGHT

DIAGONAL
SWIPE

Part II – Core Problems and Proven Fixes

(Each chapter addresses a major pain point users face.)

Chapter 3

Beating Battery Drain for Good

There's something quietly frustrating about owning a device as advanced as the iPhone 17 Pro and still watching your battery percentage fall faster than you'd expect. You charge it overnight, unplug it in the morning, and by noon it's already gasping for power. For many users, this has become the modern tech paradox—smarter phones, yet shorter stamina.

But the problem isn't always the battery itself. It's what's *silently running beneath the screen.*

This chapter uncovers those hidden power drains, breaks the myths around charging, and shows you exactly how to optimize your iPhone's energy flow—step by step, in ways that make sense for real users, including those navigating unique power realities across Africa.

Why Your iPhone Loses Power Faster Than You Think

Apple's technology is designed to do a million things for you automatically—predict your habits, refresh your favorite apps, analyze your photos, sync your messages, learn your patterns, track your health, and even prepare Siri's responses before you speak. Each of these is a micro-task that draws tiny amounts of energy every second.

The real culprit isn't one massive drain—it's **dozens of invisible ones working together.**

Let's unpack what's quietly happening:

1. **Background App Activity**

   Apps like Mail, WhatsApp, Instagram, or even Weather constantly "check in" for updates—even when you're not using them. This background fetch is convenient but relentless.

2. **Location Tracking**

Some apps use your GPS every few minutes, not because you opened them, but because they're monitoring movements for ads or analytics.

3. **iCloud Syncing**

Photos, backups, and app data upload to iCloud in the background—often when you least expect it. This is especially power-hungry during initial setup or right after taking a batch of new photos.

4. **High Refresh Rate Displays (ProMotion)**

The iPhone 17 Pro's adaptive 120Hz screen is stunningly smooth—but maintaining that fluidity consumes more energy.

5. **Thermal Load and Environment**

Warm surroundings or direct sunlight increase the battery's chemical strain, forcing it to work harder.

6. **Notification Flooding**

Every buzzing alert or glowing preview lights the display

and wakes the processor momentarily, shaving minutes off total battery life.

So when your phone feels like it's draining "for no reason," it's usually because a dozen small processes are quietly at work, each doing something helpful—but collectively exhausting.

The solution isn't to strip away convenience—it's to teach your iPhone when to rest.

The Hidden Background Refresh Killers

Background App Refresh is one of the most misunderstood features in iOS. It allows apps to stay updated even when they're closed. In moderation, it's useful—but unchecked, it's like leaving every light in your house on all night.

Here's how to tame it:

1. Audit Which Apps Are Refreshing

- Go to **Settings** › **General** › **Background App Refresh**

- You'll see a long list of apps with toggles next to them.

- Turn *off* any that don't need live data—like games, music apps, or shopping apps.

Golden Rule:

Keep Background Refresh *on* only for essentials like messaging, navigation, and email.

2. Switch to Wi-Fi Only

At the top, tap *Background App Refresh › Wi-Fi.*

This ensures that apps don't waste power refreshing over cellular data, which is far more energy-hungry.

3. Location Permissions—Fine Tune Them

Go to *Settings › Privacy & Security › Location Services*.

Change apps from *Always* to *While Using the App.*

Apps like Maps and Ride-sharing need location tracking, but others—like weather or photo filters—don't need constant GPS updates.

4. System Services You Can Disable

Under *Location Services › System Services,* toggle off:

- *Location-Based Ads*

- *Location-Based Alerts*

- *iPhone Analytics*

- *Significant Locations* (major battery drainer)

You'll still get accurate navigation, but with far less hidden activity.

Step-by-Step: Battery Health → App Usage → Optimization Settings

If your iPhone drains too fast, the *Battery* section in Settings is your truth teller. It's like a doctor's scan showing exactly where the energy goes.

1. Check Battery Health

Go to *Settings › Battery › Battery Health & Charging.*

- **Maximum Capacity:** Anything above 90% is excellent.

- **Peak Performance Capability:** Should read "Normal."

  If it says "Performance management applied," your battery may be aging.

2. Analyze Usage by App

Scroll down in *Settings › Battery.*

You'll see a graph labeled *Battery Usage by App.* Tap it.

- Apps at the top are your main power consumers.
- Check the ratio between *Screen On* and *Background Activity.*

  If Background is high, disable Background App Refresh for that app.

3. Optimize Display Power

Go to *Settings › Display & Brightness:*

- Enable *Auto-Brightness.*
- Set *Auto-Lock* to 30 seconds or 1 minute.

- Reduce brightness manually when indoors—half brightness is plenty.

4. Activate Low Power Mode When Needed

From *Control Center,* tap the yellow battery icon.

This mode temporarily pauses mail fetch, background tasks, and visual effects.

Ideal during travel or long days without access to power.

5. Review Push Notifications

Go to *Settings › Notifications.*

Turn off "Allow Notifications" for non-essential apps.

This prevents constant wake-ups that drain the battery over time.

Optimized Charging vs 100% Charge Myths

One of the most persistent myths in smartphone culture is the belief that keeping your battery at 100% is a good habit. It's not. In fact, it's one of the quietest killers of battery health.

Here's the science:

Lithium-ion batteries, like the one inside your iPhone 17 Pro, degrade faster when held at full charge or when deeply discharged.

Apple's *Optimized Battery Charging* is built to combat that.

When enabled, your iPhone learns your daily charging pattern. It charges quickly up to 80%, then pauses, holding that level until shortly before you normally unplug it (for example, when you wake up).

To enable it:

Go to *Settings › Battery › Battery Health & Charging › Optimized Battery Charging → ON*

Why this matters:

- Prevents chemical stress from constant 100% top-ups.

- Extends total lifespan by months or even years.

- Reduces heat during charging—a major cause of battery aging.

Myth vs Reality:

Myth	Reality
"I should always charge to 100%."	Aim for 80–90% for daily use.
"It's bad to plug in overnight."	Safe, if Optimized Charging is on.
"Letting my battery hit 0% recalibrates it."	Deep discharges damage the battery; recharge around 20%.

Smart Charging Habits for Nigeria's / Africa's Power Context

In many African regions, power supply can be unpredictable. That means you may rely on inverters, power banks, generators, or car charging—each with different voltage consistency. Here's how to keep your iPhone 17 Pro safe and efficient in this environment.

1. Use Certified Chargers Only

Always stick to *MFi (Made for iPhone)* certified accessories.

Cheap or unregulated adapters may provide fluctuating current, damaging your battery over time.

2. Avoid Charging Directly from Generators

If you must, use a *voltage regulator* or *UPS buffer* between your charger and generator. Unstable voltage surges can shorten battery life drastically.

3. Charging with Inverters

Keep inverter output within 220–240V AC and ensure it's pure sine wave. Avoid charging multiple high-load devices on the same inverter outlet—this causes power dips that confuse Apple's smart-charging circuit.

4. Power Banks & USB Hubs

- Choose power banks rated *20W or above (PD-certified)* for fast and safe charging.

- Avoid leaving your phone charging on low-quality USB hubs overnight; these often deliver inconsistent current.

5. Car Charging

- Always use a *dedicated lightning-to-USB-C cable* and certified car adapter.
- Never charge when your car engine is off—voltage drops during ignition can stress your phone's battery.

6. Smart Load Management

If power is intermittent, use short "top-up" charges (30–70%) rather than waiting for full cycles. This keeps your battery chemistry healthy in the long run.

The 3-Level Battery Fix Framework

(Quick Save → Deep Tune → Power Recovery)

When your iPhone battery starts acting unpredictable—draining fast, heating up, or dying suddenly—use this structured approach to restore balance.

Level 1: Quick Save (Instant Relief)

Goal: *Stop the bleeding.*

1. **Enable Low Power Mode.**
2. Turn off Background App Refresh.
3. Reduce screen brightness to 50%.
4. Disable Bluetooth and Location Services temporarily.
5. Close unused apps in the multitasking view.
6. If overheating, remove case and move to a cooler surface.

These quick adjustments can extend battery life by 30–40% immediately.

Level 2: Deep Tune (Sustained Optimization)

Goal: *Make your phone efficient long-term.*

1. Reset all Network Settings (Settings › General › Transfer or Reset iPhone › Reset › Reset Network Settings). This clears buggy connections that cause hidden background scanning.

2. Revisit App Battery Usage—delete apps with constant background activity.

3. Turn off Auto App Updates (Settings › App Store). Update manually on Wi-Fi.

4. Use Dark Mode full-time; OLED screens consume less power displaying black.

5. Keep iOS updated—battery bugs often get fixed quietly in patches.

Perform these once every few weeks for optimal performance.

Level 3: Power Recovery (Full Recalibration)

Goal: *Rebuild battery stability after months of degradation.*

1. Charge to 100%, keep plugged in for 30 minutes.

2. Use your phone normally until it hits 10%, then charge uninterrupted back to 100%.

    o This recalibrates the software's battery meter—not the battery chemistry itself, but it helps iOS estimate capacity more accurately.

3. After recalibration, re-enable Optimized Charging.

4. If battery still drains abnormally fast, consider a factory reset without restoring old backup.

    o Sometimes deep software corruption or background indexing from old backups causes persistent drain.

If all else fails and your battery health is under 80%, it's time for professional replacement.

Final Thought

Battery care isn't about restriction—it's about rhythm.

Your iPhone 17 Pro is designed to be adaptive, but it needs guidance to align with your habits. Treat your battery like a living system: let it breathe, charge it gently, avoid extremes, and keep it cool.

Once you master this chapter's framework, you'll stop watching your battery percentage with anxiety and start trusting your iPhone again—steady, responsive, and ready when you need it most.

Why Your iPhone Loses Power Faster Than You Think

Why Your iPhone Loses Power Faster Than You Think
- Background App Activity
- Location Tracking
- ICloud Syncing
- High Refresh Rate Displan
- System Services You Can Disable

‹ B... Rground App Refresh
- Background App Refresh ›
- Wi-Fi
- Calendar · Air ›
- Clock
- Instagram
- Microsoft Edge
- Outlook
- Check Battery Health
- Analyze Usage by App
- Optimize Display Power

The Hidden Background Refresh Killers

Battery Health →
App Usage →
Optimization Settings
- Battery · 86 %
- Check Battery Health
- Analyze Usage by App
- Optimize Display Power

The 3-Level Battery Fix Framework

❶ Quick Save
1. Enable Low Power Mode
2. Reduce Background Act
3. Lower Screen Brighiness

❷ Deep Tune
1. Reset Network Settings
2. Update Apps 8 IOS
3. Use Dark Mode

❸ Power Recovery
1. Recalibrate Battery
2. Factory Reset (f needed)
3. Replace Battery (if needed)

Chapter 4

Solving Overheating & Performance Lag

Your iPhone 17 Pro is a marvel of compact engineering—housing a cutting-edge A17 Bionic chip, advanced cameras, high refresh rate displays, and a cooling system designed to keep everything running efficiently. Yet, even with all this sophistication, one of the most common issues users still face is overheating.

It's unsettling when your sleek new device feels too hot to hold, starts lagging, or dims the display unexpectedly. You might think, *"This shouldn't be happening to a brand-new phone."* But it does—because heat is an unavoidable byproduct of performance. The key is not to fear it, but to understand and manage it.

In this chapter, you'll learn *why* overheating happens even in the latest iPhones, how to identify the difference between software and hardware-related heat, how to clean up hidden performance drains, and how to cool your device safely and effectively.

Why Overheating Happens Even in New Models

Even though the iPhone 17 Pro features an advanced vapor-chamber cooling system and a power-efficient 3nm A17 Bionic chip, heat remains a natural consequence of processing energy. However, what most people don't realize is that *heat alone isn't the problem—it's the imbalance between performance and power management.*

Let's look at the top causes:

1. High-Intensity Processing Tasks

4K video recording, gaming, editing large photos, or using AR (Augmented Reality) apps all push the CPU and GPU to their limits. These tasks generate heavy electrical currents through the processor, which produces heat.

2. Software Overload and Background Activity

Apps running behind the scenes—social media feeds updating,

cloud syncs uploading, widgets refreshing—can pile up unseen. These ongoing micro-processes prevent the device from resting, keeping the chip under constant low-level stress.

3. Ambient Heat and Charging Conditions

If you're charging your iPhone in direct sunlight, on a car dashboard, or under a pillow, you're essentially trapping heat. Lithium-ion batteries operate best between 0°C and 35°C (32°F–95°F). Beyond that, performance and longevity degrade quickly.

4. Network and Signal Strain

When the phone struggles to maintain a weak Wi-Fi or cellular signal, it increases its internal power draw to keep the connection stable. This leads to higher energy use and, consequently, heat.

5. Buggy Apps and Poor Optimization

Some third-party apps are not yet optimized for the latest iOS or processor architecture. They might loop background tasks, mismanage memory, or continuously trigger location services—turning into heat factories you can't easily see.

6. Overcharging and Continuous Plug-In Habits

Charging for hours beyond 100%, especially while gaming or streaming, generates heat both from the battery and power circuitry.

Even Apple's thermal system has limits—if internal temperatures exceed safety thresholds, iOS will throttle performance to prevent damage. That's when you feel lag or screen dimming, and your once-snappy phone seems to crawl.

Detecting Software vs Hardware Heat

Not all heat is created equal. To fix it effectively, you must know whether the cause lies in *software misbehavior* or *hardware stress.*

1. Software Heat — The Invisible Culprit

You'll notice:

- The phone warms up during light use (messaging, browsing, or standby).

- Certain apps trigger heat soon after launch.

- The back near the camera or the lower midsection feels warm, not the entire body.

- Performance may fluctuate but usually returns to normal after restarting.

How to Confirm:

- Go to *Settings › Battery › Battery Usage by App.* Look for apps showing unusually high background activity.

- Check *Settings › General › Background App Refresh.* Disable it for non-essential apps.

- In *Settings › Privacy › Analytics & Improvements › Analytics Data,* search for repeated crash logs of a single app. That's often your offender.

If you find one or two apps consistently at fault, delete or reinstall them. Developers often fix thermal inefficiencies in updates.

2. Hardware Heat — The Physical Factor

You'll notice:

- Heat concentrates near the bottom or side frame during charging.

- The phone gets hot even in Airplane Mode or while idle.

- The display auto-dims and frame rate drops significantly.

- Battery percentage falls sharply while the phone is warm.

This could indicate:

- Failing or aging battery.

- Faulty charging cable or adapter.

- Dust or lint in the charging port causing resistance.

- Hardware-level power IC malfunction (rare but possible).

To Confirm:

- Charge with a certified Apple or MFi charger only.

- Try a different outlet or cable.

- If the issue persists after restoring iOS and avoiding heavy use, the cause is likely hardware-related and may need professional diagnosis.

Cleaning Up Rogue Apps and Misbehaving Widgets

Even a few poorly-behaved apps or widgets can cause continuous micro-heating that drains battery life and slows performance.

Here's how to find and fix them:

Step 1: Check Recent Battery Activity

Go to *Settings › Battery.*

Under *Last 24 Hours* or *Last 10 Days,* look for apps with abnormal usage time or background percentages exceeding 10–15%.

Step 2: Disable Background Refresh and Notifications

- **Settings › General › Background App Refresh:** Turn off for apps like Facebook, TikTok, Snapchat, and shopping apps.

- **Settings › Notifications:** Turn off "Allow Notifications" for low-priority apps.

Step 3: Clear App Cache and Reinstall

Some apps, especially social and navigation ones, build massive caches that stress storage and CPU.

- Delete and reinstall them periodically (e.g., Instagram, Maps).

- Offload unused apps: Settings › iPhone Storage › Offload Unused Apps.

Step 4: Widget Detox

Widgets refresh constantly—even when you don't open them.

- Long-press on the Home Screen → tap "–" to remove widgets you rarely use.

- Limit live weather, stock, or photo widgets—they're visual but energy-heavy.

Step 5: Update or Remove Problem Apps

Always keep apps updated. Developers frequently patch performance issues.

If an app hasn't been updated in over a year or is constantly warming your device, delete it—no app is worth long-term heat damage.

Performance-Friendly Camera & Gaming Settings

The iPhone 17 Pro's power lies in its hardware—but that same power can overheat under sustained pressure. You don't have to stop using these features; you just have to use them *wisely*.

1. Camera Use Without Overheat

- Avoid long 4K or 8K recording sessions in direct sunlight.
- If filming outdoors, use a tripod shade or umbrella to reduce external heat.
- Disable *HDR video* when unnecessary (Settings › Camera › Record Video › HDR Video → Off).
- Turn off *Live Photos* if you're shooting continuously—it reduces sensor processing load.
- Don't shoot while charging. The combined battery and camera stress is a recipe for thermal buildup.

2. Gaming Without Throttling

- Lower graphics settings in games that support it.

- Close all background apps before launching heavy titles.

- Enable *Low Power Mode* before gaming—it slightly reduces peak power draw and helps stabilize frame rates.

- Avoid playing while charging; it keeps your device trapped in a constant high-heat cycle.

Bonus Tip:

Use Airplane Mode during offline gaming—it cuts heat by eliminating network background activity.

The "Cool Down Ritual" — Fast Recovery Methods

When your phone feels too warm, your instinct might be to fan it, place it near air conditioning, or even a refrigerator vent. Don't. Rapid temperature drops can cause condensation inside, damaging components.

Instead, follow this *Cool Down Ritual*—Apple-safe and effective.

87

Step 1: Pause and Unplug

Immediately stop charging and remove any case. Silicone and leather trap heat like insulation.

Step 2: Close Energy-Intensive Apps

Double-swipe up (from the bottom edge) to open the App Switcher. Close all apps actively using CPU or GPS—typically Camera, Maps, or Games.

Step 3: Enable Airplane Mode

Temporarily cut off all wireless signals. This halts network strain while still allowing you to use offline features.

Step 4: Move to a Shaded, Ventilated Space

Place the iPhone on a flat, hard surface (not fabric). Allow natural airflow—don't blow cold air directly onto it.

Step 5: Wait 5–10 Minutes

Let the temperature normalize. The device's internal sensors will regulate and restore performance automatically.

Step 6: Check for Warnings

If iOS displays a "Temperature" alert, *do not* force restart or charge. Wait until the message disappears before resuming normal use.

Once cooled, reconnect power only through a certified charger.

When to Seek Repair or Battery Replacement

Persistent heat despite all adjustments can indicate a deeper issue. iPhone batteries naturally lose capacity over time, but overheating accelerates this process dramatically.

Seek Professional Help If:

- The phone heats even when idle.
- Battery health drops below *80%* within two years.
- Charging causes instant warmth on the frame.
- Screen brightness dims randomly even in cool environments.
- Performance remains sluggish after a full software reset.

Visit an Apple Authorized Service Provider or reputable technician with diagnostic tools. They can check:

- Internal temperature logs.

- Battery cycles and resistance readings.

- Power management chip performance.

If hardware damage is confirmed, replacing the battery or logic board may be the only safe long-term fix.

Final Word

Overheating isn't always a flaw—it's your iPhone's way of protecting itself. But constant or excessive heat is a sign that something deeper needs attention.

By understanding the difference between normal warmth and destructive heat, managing your background tasks, cleaning up resource-heavy apps, and practicing the "Cool Down Ritual," you can keep your iPhone 17 Pro performing like new for years.

A cool phone isn't just more comfortable—it's faster, longer-lasting, and safer.

SOLVING OVERHEATING & PERFORMANCE LAG

WHY OVERHEATING HAPPENS

- High processing tasks
- Software overload
- Ambient heat

DETECTING SOFTWARE VS HARDWARE HEAT

SOFTWARE   HARDWARE

- Remove or reinstall apps
- Reduce widgets

CLEANING UP ROGUE APPS AND MISBEHAVING WIDGETS

- Remove or reinstall apps
- Reduce widgets

THE "COOL DOWN RITUAL"

- Pause intensive tasks
- Remove case
- Unplug and let it rest

Chapter 5

Connectivity Headaches (Wi-Fi, Bluetooth, and Mobile Data)

Your iPhone 17 Pro may be one of the smartest devices on the planet, but when your Wi-Fi drops mid-call or your Bluetooth disconnects during music playback, it can feel anything but intelligent. Connectivity problems—whether it's weak signal strength, AirDrop not working, or mobile data refusing to connect—remain some of the most common frustrations even for advanced users.

The reason isn't just "bad signal." Often, it's a mix of small misconfigurations, outdated carrier settings, or system conflicts buried deep within iOS's network layers. In this chapter, we'll take apart these connectivity headaches one by one—revealing why they happen, how to fix them safely, and how to keep your iPhone 17 Pro running smoothly across Wi-Fi, Bluetooth, and mobile data connections.

Wi-Fi Drops, Bluetooth Unpairing, and Weak Signal Causes

It's tempting to blame your carrier or router when connections fail, but the truth is: the iPhone's network systems are constantly multitasking—managing Wi-Fi, cellular, and Bluetooth simultaneously. When the system gets confused, the connections falter.

Let's break down the real causes:

1. Network Overload

When too many devices are connected to the same Wi-Fi (especially public or shared routers), your iPhone has to compete for limited bandwidth. This can cause frequent disconnects or sluggish speeds.

2. Software Bugs or Update Conflicts

New iOS updates sometimes alter how your device prioritizes

connections. A small misalignment between your router's firmware and iOS's Wi-Fi drivers can cause periodic drops.

3. Signal Interference

Other electronics—microwaves, Bluetooth speakers, or nearby routers on the same channel—can disrupt your Wi-Fi frequency. Similarly, Bluetooth devices can interfere with each other when too close.

4. iCloud and Device Sync Overload

When your device is backing up or syncing large data (Photos, Files, or iCloud Drive), the increased network load can momentarily cause slower speeds or drops.

5. Carrier and SIM Conflicts

On mobile networks, switching between 4G, 5G, and Wi-Fi can trigger interruptions, especially if APN or eSIM profiles are outdated or misconfigured.

6. Background Apps Fighting for Priority

Apps like YouTube, Spotify, or iCloud Photos can aggressively claim network resources. This hidden tug-of-war causes interruptions even when your phone is idle.

Reset Network Settings the Safe Way

If you've been troubleshooting Wi-Fi or data problems for hours with no luck, resetting your network settings can often clear the invisible junk. But it needs to be done carefully—otherwise, you'll wipe useful configurations.

Here's the *safe way* to reset:

Step 1: Back Up Your Wi-Fi Passwords

Because a reset will erase saved networks, go to *Settings › Wi-Fi › Saved Networks* (or use iCloud Keychain) and note any important passwords.

Step 2: Backup eSIM Details (if applicable)

If your carrier uses eSIM, make sure you have access to your QR code or account details. The reset doesn't usually erase eSIM profiles—but backing up is wise.

Step 3: Perform the Reset

Go to *Settings › General › Transfer or Reset iPhone › Reset › Reset Network Settings.*

Enter your passcode and confirm.

This clears:

- Wi-Fi networks and passwords

- Bluetooth pairings

- VPN and APN settings

- Network preferences

Step 4: Reboot Immediately

Restart your iPhone manually (press Volume Up $\rightarrow$ Volume Down $\rightarrow$ hold Side button). This ensures iOS rebuilds fresh network caches.

Step 5: Reconnect Strategically

- First connect to your main Wi-Fi.

- Then pair Bluetooth accessories one at a time.

- Re-enable VPNs or custom APNs last.

This systematic rebuild prevents conflicts between old and new profiles.

Diagnosing Router vs. Phone Problems

When Wi-Fi fails, the big question is: *Is it my router or my iPhone?* You can answer that scientifically in less than five minutes.

Test 1: Other Devices Check

Connect another phone or laptop to the same Wi-Fi.

- If *everything* drops, it's your router or provider.

- If *only your iPhone* struggles, the problem is local.

Test 2: Hotspot Comparison

Turn on your mobile hotspot from another phone and connect your iPhone 17 Pro to it.

- If the connection stabilizes, your iPhone hardware and iOS are fine—the issue is with your router or Wi-Fi configuration.

Test 3: Frequency Confusion

Modern routers broadcast two signals: 2.4 GHz (longer range, slower) and 5 GHz (shorter range, faster).

Your iPhone might be hopping between them.

- In router settings, give each band a unique name (e.g., *Home2.4* and *Home5G*).
- Connect your iPhone to one consistently for stability.

Test 4: Router Restart and Firmware Update

Unplug your router for 30 seconds.

Log into its admin panel (often at 192.168.0.1) and check for firmware updates. Outdated routers can't communicate properly with new iPhones using advanced encryption protocols (like WPA3).

Test 5: DNS Reset

Sometimes the problem is not your Wi-Fi, but your DNS server.

- Go to *Settings › Wi-Fi › [Your Network] › Configure DNS.*

- Select *Manual* and enter Google DNS: *8.8.8.8* or Cloudflare: *1.1.1.1.*

  You'll often notice an instant improvement.

eSIM Switching and APN Configuration for African Carriers

The iPhone 17 Pro uses dual eSIM technology—letting you run two lines digitally without physical SIMs. However, in several African regions, eSIM adoption is still maturing, and configuration issues can lead to dropped calls or no data connection.

1. Activating or Switching eSIM Profiles

- Go to *Settings › Cellular › Add eSIM.*

- Scan the QR code provided by your carrier or log in via *Carrier App* if supported (MTN, Airtel, Glo, 9mobile, etc.).

- After activation, go to *Settings › Cellular › Default Line* to set which SIM handles data, calls, and messages.

2. Troubleshooting eSIM Switch Issues

If one line shows "No Service" or "SOS Only":

- Toggle *Airplane Mode ON,* wait 10 seconds, then turn it off.
- Go to *Settings › Cellular › Cellular Data* and ensure the correct line is selected.
- Reset Network Settings (if unresolved).

3. APN Configuration for African Carriers

Some carriers don't auto-load APNs. Without them, your data won't work.

Here are examples for common networks (always confirm with your carrier):

- **MTN Nigeria:**
  - APN: *web.gprs.mtnnigeria.net*
  - Username: *web*

- Password: *web*

- **Airtel:**

  - APN: *internet.ng.airtel.com*

  - Username: *(leave blank)*

  - Password: *(leave blank)*

- **Glo Mobile:**

  - APN: *gloflat*

  - Username: *flat*

  - Password: *flat*

- **9mobile:**

  - APN: *etisalat*

  - Username: *etisalat*

  - Password: *etisalat*

To enter manually:

Go to *Settings › Cellular › Cellular Data Network*, then input your

carrier's APN.

4. Avoid Common eSIM Mistakes

- Don't delete your eSIM while troubleshooting; toggle it off instead.

- Avoid installing two eSIMs from the same carrier—they often conflict.

- Always restart after adding or switching eSIM profiles.

Smart Ways to Fix AirDrop, Hotspot, and Tethering Issues

AirDrop and hotspot functions rely on multiple connectivity layers—Wi-Fi, Bluetooth, and peer-to-peer networking. When one misbehaves, the others fall apart.

Fixing AirDrop:

1. Ensure **Wi-Fi and Bluetooth** are both ON.

2. Swipe down → *Control Center › AirDrop* → set to *Everyone for 10 Minutes* (a new privacy feature in iOS 18).

3. Make sure both devices are awake and within 30 feet.

4. If AirDrop fails repeatedly, toggle *Airplane Mode ON/OFF* to refresh local connections.

Fixing Personal Hotspot:

1. Go to *Settings › Cellular › Personal Hotspot › Allow Others to Join → ON.*

2. If it's missing, your carrier may need to enable it—contact them to activate tethering.

3. Use your *iPhone name* as the hotspot ID to avoid confusion.

4. For maximum stability, use *USB tethering* when possible—it's faster and avoids interference.

Fixing Bluetooth Pairing or Unpairing:

1. Forget the device under *Settings › Bluetooth › (i)* and re-pair it.

2. Keep Bluetooth range under 10 meters; walls and thick surfaces interfere.

3. Avoid pairing multiple audio devices at once—iOS prioritizes one audio output.

Fixing Tethering Drops (CarPlay, Modem, or PC Connection):

1. Use original Apple or MFi-certified cables only.

2. Disable *Low Power Mode* while tethering; it sometimes pauses network drivers.

3. Reset *Network Settings* if tethering repeatedly fails.

Hidden Menu: Field Test Mode for Advanced Signal Testing

If you're serious about diagnosing signal issues—especially in areas with fluctuating network coverage—Apple has a hidden diagnostic mode called *Field Test Mode.* It reveals raw signal data far more accurately than the usual "bars" icon.

To Access:

1. Open the *Phone app.*

2. Dial *3001#12345#* and press *Call.*

3. The Field Test screen will appear.

Key Readings:

- **RSRP (Reference Signal Received Power):** Measures signal strength in dBm.
  - *Good:* -70 to -90 dBm
  - *Poor:* -100 dBm or worse
- **RSRQ (Reference Signal Quality):** Measures signal clarity.
  - *Good:* -10 or higher
  - *Poor:* -15 or lower
- **SINR (Signal-to-Interference-plus-Noise Ratio):** Reflects how clean your signal is.
  - *Good:* 20 or above
  - *Poor:* below 10

If your readings are consistently poor across all tests, it's likely an environmental or carrier-side problem—not your iPhone.

Bonus Tip:

You can use apps like *OpenSignal* or *Network Cell Info Lite* to visualize coverage and tower locations near you.

Final Thoughts

Connectivity problems are rarely random—they're a conversation between your iPhone, your environment, and your network infrastructure. Once you learn how to read those signals, you gain control over every connection your phone makes.

The iPhone 17 Pro gives you the hardware; this chapter gives you the *literacy*.

By mastering Wi-Fi stability, Bluetooth pairing, eSIM setup, and mobile data optimization, you'll spend less time staring at "No Signal" and more time doing what the iPhone was built for—staying seamlessly connected, everywhere.

Wi-Fi drops, Bluetooth unpairing, and weak signal causes

- Network overload
- Software bugs
- Signal interference

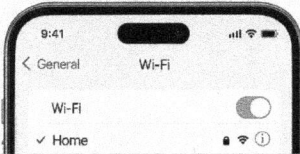

Reset Network Settings the safe way

Diagnosing router vs phone problems

- Other devices check
- Hotspot comparison

eSIM switching and APN configuration for African carriers

- Add eSIM   • Switch eSIM
- Set APN

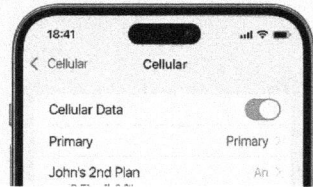

Chapter 6

App Crashes, Freezes, and Update Nightmares

Your iPhone 17 Pro is engineered to handle complex tasks effortlessly. But even the most advanced technology occasionally stumbles—apps crash, screens freeze, or updates fail halfway through. It's frustrating, especially when you're in the middle of work, a call, or editing photos.

App instability usually has less to do with your hardware and more to do with how *software, memory, and updates* interact behind the scenes. Understanding this relationship allows you to solve problems quickly—and more importantly, prevent them from recurring.

In this chapter, we'll dissect why apps misbehave, how to safely clear your device's memory, what to do when updates hang or fail, and how to simulate "Safe Mode" on an iPhone to identify rogue apps.

Why Some Apps Won't Open or Lag After Updates

You might have noticed this common pattern: you update iOS or an app, and suddenly, something breaks. Maybe Instagram won't load, your banking app freezes, or your favorite editing app keeps crashing. It's not your imagination—it's the growing complexity of the iOS ecosystem.

Here's why it happens:

1. Version Incompatibility

App developers build updates using Apple's latest developer tools. When your iPhone runs a slightly older iOS build, the app's code might call features that don't exist on your version—causing instant crashes or freezes.

2. Corrupted Cache or Data Files

Every app stores temporary data (known as cache). During an

update, those files may become incompatible with the new version. This often happens in resource-heavy apps like Facebook, TikTok, or Maps.

3. Background Sync Collisions

When apps update, iOS reindexes data like contacts, photos, and documents. If this happens while the app is running, you'll see lag, freezing, or a spinning wheel that never ends.

4. Insufficient RAM Allocation

Even though the iPhone 17 Pro has superior memory management, opening multiple high-performance apps simultaneously can exhaust active RAM. When iOS runs out of space to keep tasks "alive," it force-closes them.

5. Buggy or Unoptimized App Updates

Some app developers push new versions too quickly, without optimizing for Apple's newest chips or GPU frameworks. If this happens, you may need to wait for a bug fix.

6. Third-Party Framework Conflicts

Apps that depend on background services (e.g., VPNs, screen filters, ad-blockers) can misbehave if those services haven't been updated yet for the latest iOS release.

Pro Insight:

If multiple unrelated apps crash right after an iOS update, it's almost never the apps themselves—it's background indexing by iOS. Leave your phone plugged in overnight; by morning, it will often resolve itself.

Clearing RAM Safely on iPhone 17 Pro

The iPhone doesn't have a visible "RAM manager" like Android phones, but that doesn't mean you can't free memory when the system feels sluggish or apps lag. The process is simple, invisible, and risk-free.

Why Clear RAM?

RAM stores temporary app instructions. When it's filled with redundant data from inactive apps, your phone slows down. Clearing RAM refreshes system performance without deleting data.

How to Clear RAM Safely:

Step 1: Open *Settings › Accessibility › Touch › AssistiveTouch* and toggle it *ON*.
You'll see a small floating button appear on your screen.

Step 2: Return to *Settings › General › Shut Down*.
You'll see the "slide to power off" screen—but don't slide it.

Step 3: Tap the floating *AssistiveTouch* button, choose *Home*, and hold it until the screen flashes briefly and returns to the Home Screen.

This flushes temporary memory instantly. Apps will reload fresh when reopened, and you'll feel a noticeable boost in speed.

Alternative Shortcut:
If you prefer, simply restart your iPhone every few days. Modern

iOS handles RAM cleanup automatically, but a manual flush occasionally prevents persistent lag.

When to Force-Quit, Reinstall, or Reset

Not every app issue needs drastic measures. Here's a decision flow that helps you choose the right fix every time.

1. When to Force-Quit

If an app is frozen (unresponsive but still visible):

- Swipe up from the bottom of the screen to open the *App Switcher.*
- Swipe the app card upward to close it.
- Reopen it after 10–15 seconds.

Use this for temporary freezes or minor bugs. Don't overuse it— force-quitting frequently resets app states and actually drains more battery over time.

2. When to Reinstall

If the app keeps crashing or won't open even after restarting your phone:

1. **Delete the app** (press and hold → Remove App → Delete App).
2. **Restart your iPhone.**
3. **Reinstall** it from the App Store.

Reinstalling clears corrupted cache or outdated data that may be causing the issue.

3. When to Reset Settings

If multiple apps behave erratically or system features (like Camera, App Store, or Safari) are glitching:

Go to Settings › General › Transfer or Reset iPhone › Reset › Reset All Settings.

This resets preferences (Wi-Fi, Bluetooth, wallpaper, etc.) without erasing personal data or media. It's a clean slate for software-level conflicts.

iOS Update Stuck or Failed — Real Recovery Steps

Few things are more nerve-racking than watching your phone freeze mid-update. You're staring at the Apple logo or a progress bar that never moves. Don't panic—there's a logical way out.

Step 1: Check Internet and Power

- Ensure your iPhone is plugged in and connected to a strong Wi-Fi network. Weak connections interrupt downloads silently.

- If it's plugged into a laptop, make sure the cable isn't loose or frayed.

Step 2: Force Restart

Press and quickly release *Volume Up,* then *Volume Down*, then hold the *Side button* until the Apple logo appears.

If the update resumes, it was a temporary hang.

Step 3: Update via Finder or iTunes

If it's truly stuck, connect your iPhone to a Mac or Windows PC:

- On macOS: Open *Finder.*

- On Windows: Open iTunes.

  Your device should appear under "Locations."

  Click it and choose Check for Update.

Do NOT choose Restore yet—that wipes your data.

Finder/iTunes will download the correct iOS firmware and install it safely without data loss.

Step 4: Recovery Mode (if frozen)

If your iPhone won't boot properly:

1. Connect it to your computer.

2. Press *Volume Up,* then *Volume Down,* then hold *Side Button* until you see the "Recovery Mode" screen.

3. Finder/iTunes will detect the device and prompt you to either *Update* or *Restore*.

   o Always select *Update* first.

Step 5: Clean Restore (as last resort)

If even recovery fails, you may need to restore completely. You can restore from your iCloud or Finder backup afterward.

Using Finder / iTunes Restore Without Data Loss

Restoring is often mistaken as a "factory reset." But if done correctly, it can refresh your iOS core while preserving all personal content.

Here's how to do it safely:

1. Backup First

- In Finder, select your iPhone → click *Back Up Now*.
- Or use *Settings › [Your Name] › iCloud › iCloud Backup › Back Up Now*.

2. Enter Recovery Mode

Follow the Volume Up → Volume Down → hold Side Button sequence until the "Recovery Mode" screen appears.

3. Choose 'Update' Instead of 'Restore'

When prompted, choose *Update.* Finder/iTunes will reinstall iOS without erasing your files.

4. Wait Patiently

The process can take 15–30 minutes. Don't disconnect your phone during this time. Once completed, your iPhone will reboot with all apps and data intact.

5. Verify After Completion

Go to *Settings › General › About* to ensure your iOS version has updated. Check if the crashing apps now work properly.

"Safe Mode" Equivalent on iPhone (How to Test for Bad

Apps)

Unlike macOS, iPhones don't have a formal "Safe Mode" button. However, you can simulate one to diagnose app-related issues.

Goal:

Run iOS with only native Apple apps active to see if a third-party app is causing crashes or lag.

Method 1: Restart Without Auto-Launch Apps

1. Power off your iPhone completely.
2. Turn it back on and avoid unlocking for 2–3 minutes.
   - iOS temporarily disables certain background app refresh processes during initial boot, giving you a "clean" state to test stability.

If the phone feels fast and stable in this state, a third-party app is likely to blame.

Method 2: Use Storage Diagnostics

Go to **Settings › General › iPhone Storage.**

Scroll through the list of installed apps.

- If you see one taking up excessive storage or "System Data" abnormally high, that app might be misbehaving. Delete and reinstall it.

Method 3: Safe Testing via Reset + Gradual Reinstall

If the problem persists after major updates:

- Back up your iPhone.

- Perform a *factory reset* (Settings › Reset › Erase All Content and Settings).

- Reinstall apps *one at a time* while monitoring performance. Once lag or heat reappears, the last installed app is likely the culprit.

Pro Tips for a Crash-Free Experience

1. **Update Apps Weekly:** App updates often fix bugs faster than iOS updates.

2. **Keep Free Space:** Maintain at least 15% of total storage free; full drives cause app instability.

3. **Reboot Regularly:** Restarting every few days clears background caches.

4. **Avoid Beta Software:** Beta iOS versions are notorious for causing app instability.

5. **Use iCloud Wisely:** Offload unused apps to save resources (Settings › iPhone Storage › Offload Unused Apps).

Final Thoughts

Your iPhone 17 Pro runs one of the most stable mobile operating systems ever built—but perfection doesn't mean invincibility. Crashes, freezes, and failed updates are just signs that the system needs a reset of balance, not repair.

By mastering memory management, learning when to force-quit or reinstall, and safely restoring software through Finder, you've gained the ability to fix most issues without losing data or peace of mind.

The next time an app freezes or a progress bar stalls, remember: your iPhone isn't broken—it's simply asking for breathing room.

App Crashes, Freezes, and Update Nightmares

Why Some Apps Won't Open or Lag After Updates

- Version incompatibility
- Corrupted cache or data
- Insufficient RAM allocation

Clearing RAM Safely on iPhone 17 Pro

Settings   AssistiveTouch

Shut Down → press & hold Home

When to Force-Quit, Reinstall, or Reset

Force-Quit   Reinstall   Reset

iOS Update Stuck or Failed — Real

- Check internet and power
- Force restart

Real Recovery Steps

- Check internet and power
- Force restart
- Update via Finder / iTunes
- Enter Recovery Mode

Chapter 7

Data, Storage & iCloud Sync Issues

Storage on the iPhone 17 Pro is a bit like managing a digital home. When everything's in its place, your device runs fast, files sync smoothly, and your iCloud feels like magic. But when things get cluttered—duplicate photos, bloated apps, endless caches, or sync loops—you start to feel the weight of invisible digital chaos.

This chapter teaches you how to take back control of your data: how iCloud and local storage differ, how to free space without losing precious memories, how to keep iCloud Photos running without hiccups, and how to restore contacts or messages that mysteriously vanish.

Understanding iCloud vs Local Storage

Before you can master storage, you must understand the two systems at play: your iPhone's *local storage* and Apple's *iCloud storage.* They often work together, but they are completely separate environments.

Local storage is the physical memory inside your phone — 128GB, 256GB, or whatever model you own. Everything stored here lives *only* on your device. That includes your apps, offline photos, music, messages, and downloaded files.

iCloud storage, on the other hand, lives online — on Apple's secure servers. It's designed for syncing and backup, not for expanding your physical space. This is where your photos, notes, contacts, and app data can live in harmony across all your Apple devices.

The confusion comes when people assume iCloud adds more storage to their iPhone. It doesn't. If your iPhone is full, uploading to iCloud won't magically create more room unless you enable optimization features that offload local copies.

Think of your iPhone as your home and iCloud as your storage unit. You can keep lightweight copies in your home and store the full versions in the unit. When you need something, you simply fetch it back.

To see where your space is going, open *Settings › General › iPhone Storage.* The colored bar will show which areas (apps, photos, system, media) are consuming space. This overview is the first step in reclaiming control.

Freeing Up Space Without Losing Memories

Deleting things blindly is never the answer. The goal is to create breathing room without erasing what matters. Apple has built clever tools that let you trim the digital fat safely.

Start with Photos and Videos

If your media library takes up too much space, go to *Settings ›*

Photos › Optimize iPhone Storage. This replaces full-resolution images with smaller versions while keeping the originals in iCloud. When you tap a photo, it downloads instantly in full quality. You get the best of both worlds—accessibility and space efficiency.

Merge Duplicates

In the Photos app, scroll down to *Utilities* and open *Duplicates.* Apple's AI will find identical photos or videos and let you merge them without risk of losing detail or edits.

Clean Out Hidden Junk

Apps build caches and temporary files that grow silently over time. Safari, social media apps, and music streaming platforms are the biggest culprits.

- For Safari, go to *Settings › Safari › Clear History and Website Data.*

- For other apps, the easiest method is to delete and reinstall them. It instantly clears built-up data without touching your personal content stored in the cloud.

Review Large Files

In *Settings › General › iPhone Storage*, scroll down. You'll see a list of apps and how much space they use. Tap one to see its breakdown. If "Documents & Data" is unusually large, that's cache buildup. Reinstall the app or use "Offload App" if you want to keep its data intact.

Declutter Downloads and Messages

Messages with videos and attachments can quietly take gigabytes of space.

Go to *Settings › Messages › iPhone Storage › Review Large Attachments* and delete what you no longer need.

And remember to periodically empty the *Recently Deleted* folder in the Photos app. Items stay there for 30 days, taking up unnecessary space.

Offloading Unused Apps vs Deleting

When iOS detects low storage, it automatically suggests *Offload Unused Apps.* This is not the same as deleting.

Offloading removes the app itself but preserves all your documents, preferences, and login details. When you reinstall, it's as if nothing ever changed — everything reappears exactly as it was.

Deleting, on the other hand, removes both the app and its data. It's permanent unless that data was backed up to iCloud or within the app's own cloud service.

If you're short on space but don't want to lose progress or configurations, offload.

To do it manually, open *Settings › General › iPhone Storage*, select any app you rarely use, and tap *Offload App.*

If you're sure you'll never use it again, tap *Delete App.* Always check if that app contains data not stored elsewhere before doing so — for example, downloaded music, offline maps, or game saves.

Solving iCloud Photos Stuck Uploads and Sync Loops

One of the most common—and most frustrating—iCloud issues is seeing "Uploading…" or "Syncing…" that never ends. It can happen for dozens of reasons, from low battery to software confusion. Here's how to stop the loop and get things moving again.

Check Power and Connection

iCloud Photos pauses syncing when your phone is under 20% battery or using mobile data. Plug into power and connect to a strong Wi-Fi signal before assuming there's a bug.

Verify iCloud Space

Go to *Settings › [Your Name] › iCloud › Manage Storage.* If you're out of iCloud space, uploads will stall indefinitely. Delete old backups or unused app data, or upgrade your storage plan if needed.

Restart the Photos Sync Process

1. Go to *Settings › Photos.*

2. Turn off *iCloud Photos.*

3. Wait about thirty seconds.

4. Turn it back on.

This resets the queue without deleting your photos. Give it time—Apple's servers often take several minutes to reestablish sync.

Force a Reindex

Open *Photos* › *Albums* › *Recents* and scroll all the way down. Staying on that screen for a few minutes prompts your phone to check for pending uploads.

If Nothing Works

- Log into icloud.com and confirm your photos appear there.

- If they do, your device is likely stuck in a local indexing loop. Restart the phone, ensure it's charging, and leave it connected overnight. In most cases, sync catches up while idle.

How to Restore Missing Contacts, Messages, or Notes

Few things cause more panic than realizing your contacts or messages have disappeared. Fortunately, Apple's ecosystem is built on redundancy—most data is backed up automatically across multiple systems.

Recover Contacts

Go to icloud.com on a browser.

- Log in with your Apple ID.

- Click *Account Settings* and scroll down to *Advanced.*

- Select *Restore Contacts* and pick a version saved before they disappeared.

This replaces your current list but saves a copy of it as a backup, so you can undo the change if necessary.

Restore Messages

If you've turned on *Messages in iCloud*, toggling it back on can resync your texts.

133

Go to *Settings › [Your Name] › iCloud › Show All › Messages*, and switch it to ON. Wait a few minutes while the system syncs.

If your messages weren't synced, restore from an iCloud backup:

- Open *Settings › General › Transfer or Reset iPhone › Erase All Content and Settings.*

- During setup, choose *Restore from iCloud Backup.* Select a date before the messages went missing.

Recover Notes

In the *Notes* app, check *Recently Deleted*—you may find them there.

If not, open **Settings › Notes › Accounts.** Make sure iCloud is toggled on. Some notes might also be stored under other accounts like Gmail or Outlook, so check those folders too.

Restore from Finder or iTunes Backup

If you've previously backed up your iPhone to a computer, connect it via USB.

Open Finder (on macOS) or iTunes (on Windows).

Select your device, then choose *Restore Backup.* Pick the backup made before the loss occurred.

Your data will return exactly as it was, with messages, contacts, and notes intact.

A Smarter Way Forward

Data management isn't about deleting your life — it's about keeping it organized, efficient, and recoverable. Your iPhone is capable of handling enormous amounts of information, but only if you help it maintain balance.

Keep at least ten gigabytes of free space at all times to ensure smooth operation. Enable optimization for photos, clear app caches occasionally, and rely on iCloud as a safety net, not as extra storage.

When you understand where your data lives, syncing stops feeling mysterious and starts feeling powerful. You'll stop worrying about losing space or missing files—and start using your iPhone 17 Pro exactly as it was meant to be used: effortlessly, intelligently, and

fully in sync with your life.

Understanding ICloud vs Local Storage

☁ **iCloud storage**
- Stored in the cloud
- Sync and back up

🗑 **Local storage**
- Delete on this iPhone
- Save files and apps

▣ **Clear App Caches**
- Reinstall social meedia apps
- Clear Safari

⭐ **Review Large Attachments in Messages**
- Review Large Attachments in Messages

Offloading Unused Apps vs Deleting
- Offload: Removed in Aata
- Delete: Erased both

How to Restore Missing Contacts, Messages, or Notes

- Controls archives on iCloud com
- Recently Deleted folder in Photos or Notes
- Enable Messages in iCloud or restore a backup
- Use Restore Backup in Finder or iTunes

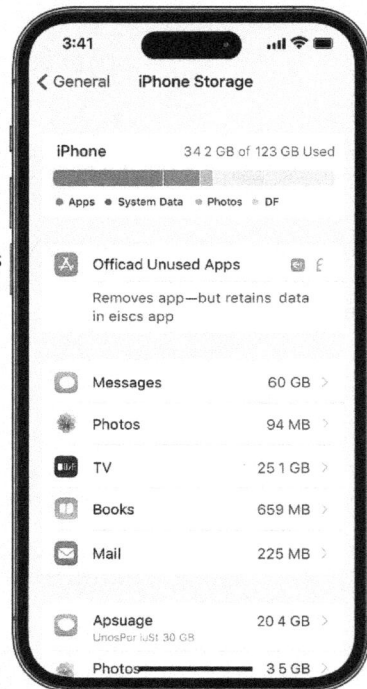

3:41   .ull 🛜 ▪

❮ General   iPhone Storage

iPhone   34 2 GB of 123 GB Used

● Apps ● System Data ● Photos ▸ DF

Officad Unused Apps
Removes app—but retains data in eiscs app

Messages   60 GB ❯
Photos   94 MB ❯
TV   25 1 GB ❯
Books   659 MB ❯
Mail   225 MB ❯
Apsuage   20 4 GB ❯
UnosPer iuSt 30 GB
Photos   3 5 GB ❯

Chapter 8

Visual, Eye-Strain & Usability Fixes

The iPhone 17 Pro is designed to be stunning — vibrant, fluid, and alive. But not everyone's eyes agree with Apple's artistic choices. The dazzling *"Liquid Glass"* visuals introduced in iOS 18, for instance, may look impressive on stage, but in everyday use, some users report discomfort, vertigo, or eye fatigue.

This chapter focuses on one of the most underappreciated aspects of iPhone mastery — *visual comfort.* You'll learn how to customize your display for your eyes, your environment, and your lifestyle. Whether you're a photographer seeking color accuracy, a senior wanting clearer text, or someone just trying to avoid eye strain during late-night scrolling, this is where your iPhone becomes truly personal.

The iOS 18 "Liquid Glass" Effect — How to Customize or

Turn It Off

The "Liquid Glass" interface in iOS 18 is Apple's latest visual evolution — a dynamic glass-like transparency that shifts subtly as you move your phone or scroll. It gives depth, realism, and a feeling of immersion. But this very fluidity can also cause sensory overload or visual fatigue for some users.

If you find yourself squinting, blinking more often, or feeling dizzy after long use, it's not your imagination — it's the *motion and contrast interplay* at work.

Here's how to tame or turn it off:

1. Reduce Transparency and Motion

- Go to *Settings › Accessibility › Display & Text Size*.
- Turn on *Reduce Transparency* to replace blurred glass backgrounds with solid colors.
- Go to *Settings › Accessibility › Motion*.

- Turn on *Reduce Motion.* This smooths or disables the parallax effects that make icons appear to "float."

2. Minimize Wallpaper Animations

Choose static wallpapers instead of dynamic ones. Go to *Settings ›*
Wallpaper › Choose a New Wallpaper, and select *Still* instead of
Dynamic.

3. Adjust Depth Effects

If you use *Lock Screen Depth Effect* (which makes elements hover over your wallpaper), turn it off for a more stable visual environment. Tap and hold your Lock Screen → tap *Customize* → disable *Depth Effect.*

4. Switch to a Matte Visual Look

Combine "Reduce Transparency" with *Dark Mode* and *True Tone* to create a neutral, low-glare screen experience that's easier on the eyes, especially in dim light.

Pro Tip:

If you love the Liquid Glass aesthetic but want less motion, you can leave *Reduce Motion* on while keeping *Transparency* off — this balances beauty and comfort.

Brightness, True Tone, and Night Shift Balancing

Brightness and color temperature play a huge role in eye comfort. The key isn't just lowering brightness — it's *balancing light dynamically* for where and when you use your iPhone.

1. Brightness Control That Adapts to You

Go to *Settings › Display & Brightness.*

- Keep *Auto-Brightness* ON (found under *Settings › Accessibility › Display & Text Size*). It uses the ambient light sensor to adjust screen levels intelligently.

- Avoid keeping your screen at 100% brightness — not only does it strain your eyes, but it also shortens battery life and increases heat.

140

2. True Tone for Real-World Lighting

True Tone automatically adjusts white balance to match your environment. Under warm indoor light, your screen becomes slightly yellow; under daylight, slightly cooler.

Enable it under *Settings › Display & Brightness › True Tone.*

It mimics how paper looks under natural lighting, reducing the harsh blue cast that causes fatigue.

3. Night Shift — Your After-Hours Ally

Night Shift shifts your screen's color temperature toward warmer tones after sunset.

- Go to *Settings › Display & Brightness › Night Shift.*
- Set *Schedule* to *Sunset to Sunrise.*
- Adjust the *Color Temperature* slider slightly to the right for "More Warm."

Warm tones reduce blue light exposure, which disrupts sleep by tricking your brain into thinking it's still daytime.

4. The Manual Brightness Trick for Night Readers

If you read in total darkness, enable *Reduce White Point* under *Settings › Accessibility › Display & Text Size.* Lower it to around 30%.

This dims bright whites without affecting overall brightness — perfect for bedtime reading without hurting your eyes.

Font Size & Bold Text for Seniors

A beautiful display means little if it's hard to read. As screen resolutions improve, text gets sharper—but also smaller. Apple includes powerful readability tools for anyone who prefers larger, clearer text.

1. Adjust System Font Size

- Go to *Settings › Display & Brightness › Text Size.*
- Drag the slider to increase text size across all system apps.

For even larger options, enable *Settings › Accessibility › Display & Text Size › Larger Text.*

Turn on *Larger Accessibility Sizes* for extreme readability, great for seniors or users with visual impairments.

2. Enable Bold Text

In *Settings › Accessibility › Display & Text Size,* toggle *Bold Text.* It enhances contrast, making words stand out sharply against backgrounds.

3. Use Display Zoom (Interface Scaling)

Go to *Settings › Display & Brightness › Display Zoom › Larger Text.* This slightly enlarges icons, buttons, and text across the interface — not just fonts. It's ideal for those who prefer a less cluttered layout.

4. Test Legibility in Your Most-Used Apps

Open Messages, Mail, and Safari after adjusting fonts. If certain text still feels small, use *Reader View* in Safari (tap the *aA* icon in the address bar) for a clutter-free, large-font reading experience.

Custom Display Profiles for Photographers and Night

Readers

The iPhone 17 Pro's OLED Super Retina XDR display is built for precision. But depending on your use case—whether it's photo editing, content creation, or relaxing at night—you can create a tailored display feel without sacrificing color accuracy.

1. For Photographers and Designers

Color accuracy is everything when editing images or comparing tones.

- Disable *True Tone* for consistent color temperature.

- Turn off *Night Shift* to maintain neutral whites.

- Use *Light Mode* (Settings › Display & Brightness) for natural color preview.

- Adjust brightness manually to about 70% to match average monitor brightness.

If you use third-party editing apps like Lightroom or VSCO, calibrate your edits in stable lighting. The iPhone's OLED can exaggerate contrast in darker rooms, so check your photos in daylight before final export.

2. For Night Readers and Writers

Comfort is the goal here, not accuracy.

- Enable *Dark Mode.*

- Activate *Night Shift* with "More Warm" selected.

- Reduce *White Point* to around 40%.

- Combine with *Bold Text* and slightly larger fonts for sustained reading comfort.

This setup gives a paper-like experience that's gentle on the eyes, perfect for bedtime use or reading long documents.

3. For Outdoor or Daylight Users

- Keep *True Tone* ON.

- Turn on *Auto-Brightness.*

- Use *Light Mode* for maximum visibility under sunlight.

- Disable *Reduce White Point*—it can make the screen too dim outdoors.

Accessibility Shortcuts You'll Actually Use Daily

Accessibility isn't just for people with disabilities—it's about efficiency and comfort. Apple's accessibility suite is full of hidden gems that make daily use faster and easier.

1. Enable the Accessibility Shortcut

Go to *Settings › Accessibility › Accessibility Shortcut.*

This lets you assign up to three quick actions triggered by triple-pressing the Side Button.

Recommended shortcuts:

- *Reduce Motion* (to calm dynamic effects)
- *Zoom* (for quick text enlargement)
- *Color Filters* (for glare reduction or color sensitivity)
- *Voice Control* (for hands-free navigation)

2. Create a Quick Visibility Preset

Combine *Reduce Motion*, *Dark Mode*, and *Night Shift* into one shortcut. You can switch between "Day Mode" and "Comfort Mode" in seconds.

3. Use Back Tap for Instant Actions

Go to *Settings › Accessibility › Touch › Back Tap.*

Assign actions like *Open Control Center*, *Take Screenshot*, or *Toggle Dark Mode* to a double or triple tap on the back of your phone.

4. Use "Focus Filters" for Visual Peace

Within *Focus Modes,* you can hide distracting home screens or dim notifications during reading, work, or rest. It's a visual decluttering trick that also improves concentration.

A Healthier Way to See Your Screen

The iPhone 17 Pro's screen is not just a window into your digital life—it's a light source you stare at for hours each day. Managing it properly isn't vanity; it's self-care.

By customizing motion, brightness, and font size, you reduce fatigue, protect your eyes, and make your phone feel truly *yours*. The right setup doesn't just look better—it *feels* better.

Your iPhone should never demand that you adapt to it; it should adapt to you. And now, with the tools you've unlocked, it will.

IOS 18 "Liquid Glass" Effect

- Reduce transparency and motion

- Minimize wallpaper animations

- Depth Effect (Lock Screen)

Brightness, True Tone and Night Shift

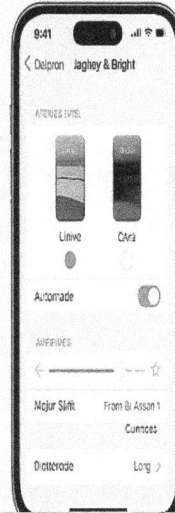

- Enable Auto-Brightness

- Use True Tone to match ambient light

- Schedule a warm screen at night

Font Size & Bold Text for Seniors

- Adjust system font size

- Turn on Bold Text

Accessibility Shortcuts You'll Actually Use Daily

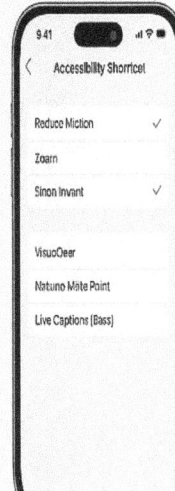

- Assign one or more settings

- Triple-click the Side button

- Try Reduce Motion, Zoom

Part III – Everyday Mastery & Hidden Tricks

(Turn frustration into joy and mastery.)

Chapter 9

Camera & Photo Magic

Few things define the iPhone more than its camera. For many users, it's not just a tool for memories—it's a professional instrument for storytelling, content creation, and even business. The iPhone 17 Pro's camera system is the culmination of Apple's most advanced engineering yet: triple lenses, massive sensors, ProRAW image capture, cinematic video control, and processing so intelligent it can edit before you even open your gallery.

But to truly unlock its potential, you need to go beyond the basics. Apple hides some of its most powerful features beneath layers of menus and icons—features that can transform how you shoot and edit. In this chapter, we'll uncover those hidden settings, simplify the complex shooting modes, teach you how to manage heat during heavy video sessions, and show you how to build a creator-grade storage workflow that keeps your projects organized and efficient.

Hidden Camera Settings Apple Doesn't Explain

Open your Camera app, and it looks simple. But hidden behind that simplicity are powerful professional-grade controls. The trick is knowing where to look.

1. Preserve Camera Settings

By default, your iPhone resets certain settings each time you close the Camera app. That means if you prefer ProRAW, manual exposure, or specific filters, you'll have to reapply them every time—unless you change this.

Go to *Settings › Camera › Preserve Settings.*

Enable options like *Camera Mode*, *Exposure Adjustment*, *Live Photo*, and *Creative Controls.*

Now, every time you reopen the Camera, it remembers your last setup—saving precious seconds when capturing fast-moving scenes.

152

2. Turn on Grid and Level

Go to *Settings › Camera › Grid.*

This activates the rule-of-thirds grid for better composition and adds a horizon level indicator when shooting flat surfaces—essential for architectural and landscape photography.

3. Hidden Resolution Controls

In iOS 18, you can switch between resolutions directly from the viewfinder. Tap the small "RAW/HEIF/JPG" or "4K/HD" indicators at the top to toggle formats instantly without diving into Settings.

4. Macro Control Toggle

The iPhone 17 Pro automatically switches to macro mode when close to subjects, but it can be unpredictable.

Go to *Settings › Camera › Macro Control* and turn it ON.

Now you'll get a small flower icon that appears when you're close to a subject—you can tap it to manually enable or disable macro focus.

5. View Outside the Frame

This subtle but powerful feature helps with composition.

In *Settings › Camera › View Outside the Frame,* enable it to preview what's just outside your shot boundaries. It's invaluable for planning framing or anticipating movement.

6. Hidden Focus Peaking and Zebra Pattern (for Video Pros)

If you record videos manually using third-party apps like *FiLMiC Pro* or Apple's *Pro Camera mode*, you can enable Focus Peaking (highlights sharp areas) and Zebra Patterns (shows overexposed regions).

These are hidden in *Settings › Camera › Formats › Pro Video Tools.*

ProRAW, Depth Control, Night Mode, and Action Mode Simplified

Apple's shooting modes can sound intimidating, but each one serves a clear purpose. Once you understand them, you'll know exactly when to use each—and why.

ProRAW — For Professionals and Perfectionists

ProRAW combines Apple's computational photography (Smart HDR, Deep Fusion) with the flexibility of a RAW file. It's perfect for serious edits, as it retains detail and dynamic range far beyond JPEG or HEIF.

When to Use:

- Low-light scenes where you want to recover shadows.
- Portraits where color grading or skin tone correction is planned.
- Product or landscape photography where color accuracy matters.

How to Enable:

Go to *Settings › Camera › Formats › Apple ProRAW* and select 48MP.

In the Camera app, tap *RAW* at the top of the viewfinder to toggle it.

155

Depth Control — Perfect Portraits Made Simple

Depth Control lets you adjust the blur behind your subject after shooting. It's available for Portrait Mode photos.

Open a Portrait photo in *Photos*, tap *Edit*, then tap the "f/number" icon. Drag the slider—lower numbers (f/1.4–f/2.8) increase blur, higher numbers reduce it.

When to Use:

- For professional headshots or product photos.
- To separate your subject from distracting backgrounds.

Night Mode — Low Light, Bright Detail

Night Mode uses multi-frame processing to capture clean, bright photos in darkness. It automatically activates in dim conditions, showing a yellow moon icon.

Tap it to adjust exposure duration—shorter times (1–2 seconds) prevent blur, while longer times (up to 10 seconds) reveal deep color and texture.

Stabilize your phone on a surface or tripod for best results.

156

Action Mode — For Motion Without Blur

Action Mode uses advanced stabilization to keep videos smooth even when you're running or panning quickly.

Tap the *running figure* icon in Video mode to enable it. It reduces shake dramatically but uses extra light, so it's best outdoors or in bright environments.

Heat Management During Long 4K Recording

Shooting in 4K—or worse, 4K ProRes—turns your iPhone into a miniature film studio. But all that power generates heat. If your phone gets too hot, iOS automatically lowers brightness, reduces frame rate, or even stops recording.

To prevent this, you need to manage both *internal heat* and *external conditions.*

1. Avoid Charging While Recording

Recording while plugged in doubles heat output. Always start with a full battery, or use a high-quality MagSafe power bank that doesn't trap heat directly against the phone's back.

2. Use Airplane Mode During Filming

Turning off Wi-Fi, Bluetooth, and 5G eliminates background radio activity, freeing up processing power for video capture and keeping your device cooler.

3. Record in Short Sessions

Break long shoots into shorter clips (5–10 minutes). This not only prevents overheating but also makes editing and syncing easier later.

4. Lower Recording Settings When Possible

If your project doesn't require extreme detail, record at 1080p 60fps or 4K 30fps instead of 4K 60fps. The quality difference is minimal for most screens, but heat reduction is significant.

5. Remove the Case

Even a thin silicone case traps heat. Remove it during long recordings to improve thermal dissipation.

6. Use External Cooling

For professional filming, consider a small USB-powered cooling fan or metal heat sink attachment. These are compact, silent, and can reduce overheating by up to 40%.

Pro Tip:

If your iPhone dims during recording, it's a warning sign that internal temperatures are climbing. Stop recording and let it rest in shade for 3–5 minutes before resuming.

Smart Storage Workflow for Creators

The iPhone 17 Pro produces enormous files, especially in 48MP ProRAW or 4K ProRes video. Without an organized system, you'll quickly drown in gigabytes. Here's how to build a professional workflow that keeps everything safe and manageable.

1. Record Directly to External SSD (Pro Workflow Only)

Connect a USB-C SSD (like a Samsung T7 or SanDisk Extreme) using a short, high-speed cable.

When you open the Camera app, you'll see *External Storage* appear automatically. You can record ProRes directly to it — freeing up internal space instantly.

2. Use iCloud for Active Projects

Keep iCloud Photos enabled for ongoing work. It's ideal for syncing between your iPhone, iPad, and Mac without manual transfers.

But don't rely on iCloud for archiving finished projects—it's meant for synchronization, not long-term storage.

3. Handoff Editing to Mac Seamlessly

With *AirDrop* or *Handoff*, you can pick up edits right where you left off.

- Open your photo or video on the iPhone.
- Tap *Share › AirDrop* and choose your Mac.

  Once received, it opens instantly in *Photos* or *Final Cut Pro* for desktop-level editing.

4. Archive Long-Term Projects Externally

Once you've completed a project, move it to:

- **An external SSD or HDD** for physical backup.

- **iCloud Drive or Google Drive** for cloud redundancy.

To export efficiently, use the *Files* app to drag items from your camera roll directly into your connected drive. This bypasses compression that can occur in AirDrop or messaging apps.

5. Maintain a Three-Tier System

- **iPhone:** Current work only.

- **Cloud:** Projects in progress.

- **External Drive:** Final archives and backups.

Following this cycle keeps your phone lean, your data secure, and your creative flow uninterrupted.

Final Thoughts

The iPhone 17 Pro's camera is no longer just a mobile tool — it's a full-fledged creative powerhouse. But true mastery isn't about pressing "record"; it's about knowing when to use the right format, how to manage the heat of a shoot, and how to keep your storage running like a professional studio.

Once you understand these hidden mechanics, your iPhone transforms from a smartphone into a pocket cinematographer's dream — one capable of producing content fit for global screens, right from your hand.

HIDDEN CAMERA SETTINGS APPLE DOESN'T EXPLAIN

- Preserve Camera Settings
- Grid and Level
- Macro Control Toggle
- View Outside the Frame

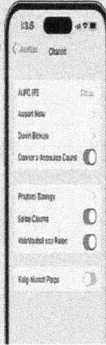

PRORAW, DEPTH CONTROL, NIGHT MODE, AND ACTION MOD SIMPLIFIED

HEAT MANAGEMENT DURING LONG 4K RECORDING

- Avold charging while recording
- Use Airplane Mode during filming
- Record in short sessions
- Lower recording settings when possible

SMART STORAGE WORKFLOW FOR CREATORS

- Record directly to external SSD
- Use iCloud for active projects
- Handoff editing to Mac seamlessly
- Archive long-term projects externally

Chapter 10

Communication Without Glitches

In the digital world, communication isn't just about messages — it's about reliability. When your iPhone 17 Pro refuses to send an iMessage, FaceTime won't activate, or your WhatsApp behaves strangely with two SIM cards, it can feel as if the smartest phone in the world suddenly forgot how to talk.

This chapter restores that confidence. You'll learn how to solve stubborn iMessage and FaceTime activation issues, manage multiple SIM lines and chat apps smoothly, configure your email for clean syncing, and strengthen your privacy without losing convenience. Once you master these, your iPhone becomes what it was meant to be: a device that keeps you connected seamlessly, anywhere in the world.

Fixing iMessage and FaceTime Activation Errors

It's one of the most common—and most irritating—issues: you try to activate iMessage or FaceTime and see "Waiting for Activation…" for hours, sometimes days. Don't worry—this is fixable.

Both services rely on Apple's secure servers verifying your Apple ID and phone number. If something interrupts that handshake, activation fails. Let's fix that step by step.

1. Check Your Network and Time Settings

- Make sure your iPhone has a strong *Wi-Fi or mobile data connection.*

- Go to *Settings › General › Date & Time* and ensure *Set Automatically* is ON. Incorrect time zones often block activation.

2. Verify Apple ID and Number

- Go to *Settings › Messages › Send & Receive.*

- Ensure your correct Apple ID and phone number are checked.

- Do the same for *Settings › FaceTime.*

3. Toggle iMessage and FaceTime Off and On

- Turn both off, wait 30 seconds, and turn them back on.

- You may receive a message that "Your carrier may charge for SMS." Tap OK — that's normal, as Apple uses a one-time text for verification.

4. Sign Out and Back Into iCloud

- Go to *Settings › [Your Name] › Sign Out.*

- Restart your phone, then sign back in and retry activation.

5. Check for Carrier Restrictions

If you're using a dual SIM or eSIM setup, ensure your primary line supports SMS and international texting. iMessage activation requires it. If one line is data-only, switch temporarily to the one with messaging enabled.

6. Reset Network Settings (if still stuck)

Go to *Settings › General › Transfer or Reset iPhone › Reset › Reset Network Settings.*

This refreshes all communication protocols without deleting personal data.

7. Final Option — Use Apple's Servers via VPN or Different Network

Sometimes, your carrier's routing blocks Apple's activation servers. Connect to a trusted VPN or different Wi-Fi network and retry.

Once activated, your messages and FaceTime calls should instantly sync across all Apple devices signed in with the same ID.

Dual SIM and WhatsApp Management

The iPhone 17 Pro supports *dual eSIMs **or** one physical SIM plus one eSIM,* allowing you to manage personal and work numbers on the same device. It's powerful, but it can get messy with messaging apps like WhatsApp.

1. Setting Up Dual SIMs Correctly

- Go to *Settings › Cellular › Add eSIM.*

- Activate your second line using a QR code from your carrier.

- Label each line (e.g., *Work* and *Personal*). This helps later when sending messages or calls.

2. Choosing Which Line Handles Data

Under *Settings › Cellular › Cellular Data*, pick one line for data. The other can still receive calls and SMS.

3. Using WhatsApp With Dual SIM

WhatsApp officially supports one number per app. However, you can run *two WhatsApp accounts* on your iPhone:

- Install the main WhatsApp app for one number.

- Install *WhatsApp Business* from the App Store for the second number.

Each app verifies independently using its assigned line. It's the cleanest, safest dual-account setup without violating WhatsApp's policies.

4. Managing iMessage With Dual Lines

iMessage can work with both numbers, but you must select which one to use for outgoing messages.

Go to *Settings › Messages › Send & Receive › Start New Conversations From,* and choose your preferred number.

5. Smart Tip for Travelers

If you often switch regions, use your eSIM for local data plans while keeping your home number active for calls. This saves on roaming and keeps communication consistent across borders.

Email Syncing and Notification Best Practices

Your iPhone 17 Pro's Mail app can sync multiple accounts effortlessly — but poor configuration can cause missing emails, duplicate alerts, or excessive battery drain. To stay organized and efficient, you need to fine-tune your email system.

1. Use "Fetch" Instead of "Push" for Non-Essential Accounts

Go to **Settings › Mail › Accounts › Fetch New Data.**

- Keep *Push* on for your main work or personal account (for instant delivery).
- Set other accounts to *Fetch Hourly* or *Manually.*

This single adjustment can dramatically improve battery life.

2. Delete and Re-Add Glitchy Accounts

If emails stop syncing, remove the account and add it back.

- Go to **Settings › Mail › Accounts › [Account Name] › Delete Account.**
- Restart your iPhone and re-add it.

This clears corrupted sync tokens that occasionally get stuck after iOS updates.

3. Organize Notifications Intelligently

- In *Settings › Notifications › Mail,* disable notifications for unimportant accounts.

- For essential ones, choose *Time Sensitive* or *Critical Alerts* only.

  This helps you stay informed without being constantly interrupted.

4. Enable Threading and VIP Lists

In *Settings › Mail › Threading,* enable *Organize by Thread* to group conversations neatly.

Use *VIPs* to highlight messages from key contacts—they'll show up separately and can trigger custom notifications.

5. Use the "Swipe Gestures" for Productivity

In *Settings › Mail › Swipe Options,* customize your swipes.

For example, assign "Swipe Left" to *Flag* and "Swipe Right" to *Trash* or *Archive.*

This keeps your inbox clean in seconds.

171

6. Avoid Heavy Attachments in iCloud Mail

Large attachments in iCloud Mail sync slowly across devices. Instead, upload large files to *iCloud Drive* and share via a link. It's faster, cleaner, and prevents sync delays.

Privacy Tweaks That Prevent Unwanted Data Use

Apple's ecosystem is designed to be secure by default, but there are still quiet ways apps and services collect information. A few simple settings can keep your communication private without sacrificing performance.

1. Limit Message and Email Tracking

Go to *Settings › Privacy & Security › Mail Privacy Protection,* and turn it ON.

This hides your IP address and blocks invisible tracking pixels in emails that reveal when you've read them.

2. Disable Link Tracking in Messages

iOS 18 introduces automatic *link tracking protection,* but verify it's on:

Settings › Privacy › Messages › Link Tracking Protection → ON.

This removes embedded tracking codes from URLs before opening them.

3. Control App Access to Contacts and Messages

Go to *Settings › Privacy & Security › Contacts.*

Review which apps have access. Disable any that don't need it— especially social media or unknown apps.

4. Limit Siri Data Collection

If you use Siri often, Apple still collects anonymized samples for improvement.

To disable this: *Settings › Siri & Search › Siri Analytics & Improvements → OFF.*

You'll still enjoy Siri's convenience without contributing voice data to analytics.

5. Block Unwanted Caller and Message Tracking

173

- Turn on *Silence Unknown Callers* under *Settings › Phone.*

- In *Messages › Unknown & Spam,* enable *Filter Unknown Senders.*

  These features create a digital shield, keeping your focus on real conversations.

6. Review Location Sharing in Messages and FaceTime

In *Settings › Privacy & Security › Location Services,* scroll to *Share My Location.*

Only allow people you trust — and review this list occasionally.

7. Hide Your Email for Safer Signups

Use *Hide My Email* (found under *Settings › [Your Name] › iCloud › Hide My Email*) when signing up for newsletters or unfamiliar services. It creates unique, disposable addresses that forward to your real inbox, protecting your identity.

Final Thoughts

The power of the iPhone 17 Pro isn't just in its hardware—it's in how effortlessly it connects you to the world. But that ease of connection must come with control.

When you understand how to stabilize iMessage, manage dual SIMs, balance email notifications, and guard your privacy, your phone becomes more than a device—it becomes a secure communication hub you can rely on, no matter where you are.

Fixing iMessage and FaceTime Activation Errors

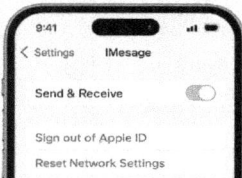

9:41
< Settings   IMesage
Send & Receive
Sign out of Apple ID
Reset Network Settings

- Sign out of Apple ID
- Reset Network Settnings
- Use VPN

Dual SIM and WhatsApp Management

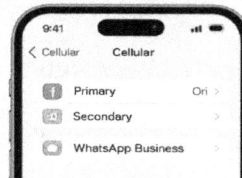

9:41
< Cellular   Cellular
Primary   Ori >
Secondary   >
WhatsApp Business

- Add an e-SIM lines
- Label lines "Work" and "Personal"
- Llabel lines "Work" Personal"
- Use VPN

Email Syncing and Notification Best Practiices

9:41
< General   Fetch New Data
Push
iCloud   Airomalic >
Gmail   Fetch >
Yahoo!   Every 15 Minutes

- Push only for essential accounts
- Delete and e-add accounts
- Avaid email notifaications don't

Privacy Tweaks That Prevent Unwanted Data Use

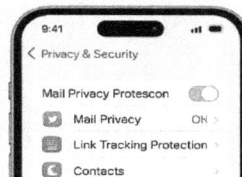

9:41
< Privacy & Security
Mail Privacy Protescon
Mail Privacy   ON >
Link Tracking Protection
Contacts

- Enable Mall Privacy Protection
- Tum on Link Tracking
- Review app access to Contacts

175

Chapter 11

Personalization Without Breaking the System

Apple designed the iPhone 17 Pro to feel personal right out of the box—but for many users, personalization is more than just wallpaper and widgets. It's about creating a phone that feels *uniquely yours* while staying fast, stable, and efficient. Unfortunately, that's where most people go wrong: too many widgets, poorly designed shortcuts, or unstable third-party themes can quietly cripple performance and battery life.

This chapter is your blueprint for smart customization—one that blends creativity with reliability. You'll learn how to build a beautiful, functional interface, automate your day with Focus Modes and Shortcuts, and experiment safely without risking system crashes or data loss.

Widgets That Don't Drain Power

Widgets can transform your iPhone's Home Screen from static to smart—giving you live weather, reminders, battery info, or photos at a glance. But not all widgets are equal. Some update constantly, using location and network access that can chew through battery life.

To personalize intelligently, think *lightweight, not flashy.*

1. Choose Wisely: Use Apple's Native Widgets First

Apple's built-in widgets (like Calendar, Weather, Reminders, Batteries, and Screen Time) are optimized for iOS efficiency. They refresh less often and integrate directly with system services. Third-party widgets, especially those tied to social media or news feeds, update far more frequently and use background refresh power.

2. Limit Active Widgets per Screen

Each widget consumes a slice of your system's live memory. The more dynamic widgets you have, the more the iPhone must redraw data. Keep only what you check daily—like weather, calendar, or battery level. One or two widget stacks per screen are ideal.

3. Use "Smart Stacks" Instead of Multiple Single Widgets

A Smart Stack automatically rotates relevant widgets based on time or context (e.g., Calendar in the morning, Fitness mid-day, Photos in the evening).

To create one:

- Touch and hold your Home Screen.

- Tap the "+" in the corner, select *Smart Stack*, and add your preferred size.

- You can also drag other widgets into an existing stack.

This method gives variety without constant background refreshes.

4. Disable Widget Location Access When Unnecessary

Widgets like Weather or Maps use GPS frequently. Go to Settings › Privacy & Security › Location Services, scroll to the widget's app, and choose *While Using* or *Never*. This prevents silent background tracking.

5. Avoid Animation-Heavy Widgets

Some creative widgets use dynamic animations or custom refresh intervals—these look great but drain battery fast. Prioritize static or system-optimized visuals.

Focus Modes, Shortcuts Automation, and Home Screen Themes

The magic of iOS 18 lies in its ability to adapt to your lifestyle. Focus Modes and Shortcuts automation turn your phone from a distraction machine into a personal assistant—if you set them up properly.

1. Focus Modes — The Art of Controlled Connection

Focus Modes let you decide *when* and *how* your phone notifies you. Perfect for work, sleep, study, or even creative hours.

To set up:

- Go to **Settings › Focus.**

- Choose or create a mode (e.g., Work, Sleep, Personal).

- Select people or apps allowed to notify you during that mode.

Pro Tip:

Use "Focus Filters" to adjust system behavior automatically. For example:

- Hide personal mail during work.

- Limit Safari to specific tabs.

- Enable Dark Mode and reduce brightness during reading hours.

Once configured, you can switch Focus modes manually or trigger them with automation (see below).

2. Shortcuts Automation — Your Digital Routine Designer

Shortcuts turn multi-step actions into single taps or automated events. Imagine this: your phone switches to Silent Mode, lowers brightness, and opens Apple Books automatically when you plug in your charger at night. That's the power of automation.

To get started:

- Open *Shortcuts App › Automation › New Automation.*

- Choose a trigger (e.g., Time of Day, App Launch, or Charging).

- Add actions: control settings, send messages, or launch apps.

- Save and toggle *Ask Before Running* to OFF for full automation.

Examples of Useful Shortcuts:

- **Morning Routine:** When alarm stops → Open Weather + Calendar + Apple Music.

- **Work Mode:** At 9:00 a.m. → Enable Focus Mode + Reduce Notifications + Open Notes.

- **Battery Saver:** When battery < 20% → Turn Off Bluetooth, Reduce Brightness, Enable Low Power Mode.

Pro Tip:

Keep automations minimal. Too many simultaneous triggers can cause lag or background drain. Prioritize 3–5 well-designed automations for maximum benefit.

3. Home Screen Themes — Balanced Aesthetics, Not Overload

Apple now allows deep customization of icons, fonts, and themes using the Shortcuts app and custom wallpapers. But many users fall into a trap—creating hundreds of new app icons that slow down load times.

Here's how to create a clean, efficient theme without breaking stability:

- **Use Folders Intelligently:** Group rarely used apps into folders on secondary screens. Keep only essentials (Camera, Messages, Safari, Notes) on your primary page.

- **Customize with Restraint:** Use consistent icon styles and colors. Overly mixed themes confuse your eye and drain focus.

- **Use Custom Icons Properly:** When creating custom icons via Shortcuts, toggle "Open Immediately" and hide the Shortcut notifications. Otherwise, you'll see a flash every time you open an app.

- **Leverage Wallpapers:** Choose calm, solid backgrounds that enhance readability. Avoid bright neon or busy images that strain the eyes or make text hard to read.

If you love dynamic visuals, enable *Depth Effect Wallpapers*—they add depth without animation-heavy drain.

Balancing Style vs Stability

Personalization should never come at the cost of usability. If your iPhone starts lagging, heating, or draining battery unusually fast after heavy customization, it's time to scale back.

1. Use Minimal Animations

Disable unnecessary motion effects: **Settings** › **Accessibility** ›

Motion › Reduce Motion → ON. This keeps transitions smooth but efficient.

2. Limit Live Wallpapers

Live wallpapers constantly refresh background visuals—even when the screen is off briefly. Use still or dark wallpapers instead for better battery and reduced GPU load.

3. Avoid Third-Party "Launcher" Apps

Some apps promise advanced themes or icon packs but hijack system behavior. They can disrupt shortcuts, overload background tasks, and even cause crashes. Stick to Apple's native tools and trusted apps from verified developers.

4. Monitor Battery Impact

Go to **Settings › Battery › Battery Usage by App.**

If you notice unusual drain from the *Shortcuts* or *Widgets* apps, simplify your layout. Custom automation loops sometimes get stuck.

5. Test Before Expanding Customization

Change one design or automation at a time and monitor your phone's behavior for 24 hours. This way, if something goes wrong, you'll know exactly what caused it.

Backup Before Experimenting with Customization

Before diving into any deep customization—new themes, layout overhauls, or automation workflows—create a backup. Customization mistakes can cause icon misalignment, stuck shortcuts, or even UI crashes after iOS updates.

1. Use iCloud Backup

Go to *Settings › [Your Name] › iCloud › iCloud Backup › Back Up Now.*

This saves your apps, data, and layout preferences automatically.

2. Create an Encrypted Backup on Your Mac

Connect your iPhone via USB → open Finder → select your device → check *Encrypt Local Backup* → click *Back Up Now.*

Encrypted backups preserve Wi-Fi settings, health data, and shortcuts automations.

3. Keep Screenshots of Your Layout

Before making major theme changes, take screenshots of your Home Screen layout. It's a quick visual reference in case you need to rebuild manually.

4. Store Custom Icons and Wallpapers in iCloud Drive

Save all icon sets, color palettes, and wallpapers in one iCloud folder labeled *Theme Resources.* This ensures that if your layout resets after an update, you can reapply your theme quickly.

5. Don't Forget to Reboot

After big layout or automation changes, restart your iPhone. It clears cache and refreshes background indexing—preventing small glitches from turning into performance problems.

Final Thoughts

Personalization is a delicate dance between expression and efficiency. Done right, your iPhone 17 Pro becomes a reflection of your habits and identity—organized, stylish, and stable. Done carelessly, it becomes cluttered, slow, and frustrating.

The best design philosophy? *Intentional simplicity.* Every widget, automation, and wallpaper should serve a purpose—making your daily experience easier, not heavier.

Remember: creativity should enhance usability, not fight it.
Back up your work, customize smartly, and enjoy a phone that truly feels like *you*—powerful, personal, and perfectly balanced.

Personalization Without Breaking the System

MON
15
Call Anna

Work
75°
Pretty Cloudy

100%

Back Up Now
Do you noitr, to beck like data.
un atitudds proo and oitis on the
tho use / I wers sponation placem
towsnfs the iPhone i I slill-lh and
on vit #1 it laps moliciually.

Cancel   Back Up Now

Focus
‹ Focus

Work ›
Personal ›
Sleep ›
Do Not Disturb ›

Home Screen Themes

**Backup before
experimenting wit
customization**

Widgets that don't
drain power

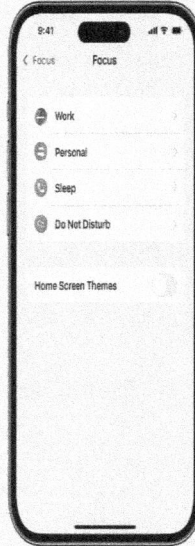

Focus Modes, Shortcuts
automation, and Home Screen themes

Chapter 12

Security, Privacy & Recovery

Your iPhone 17 Pro isn't just a device—it's the most personal piece of technology you own. It holds your photos, conversations, passwords, bank details, location history, and daily routines. Losing control of it—whether through theft, hacking, or careless disposal—can feel like losing a part of yourself.

Apple built iOS with some of the strongest consumer privacy protections in the world, but they only work when you use them fully. This chapter shows you how to protect your iPhone against theft and scams, secure your identity with Face ID and passkeys, lock your iCloud behind two-factor authentication, track a lost phone even when it's turned off, and safely wipe it before selling or giving it away.

Protecting Your Device from Theft and Scam Calls

Modern thieves don't just steal phones—they steal identities. Protecting your iPhone means thinking beyond physical safety to digital resilience.

1. Use a Strong Passcode (Not Just Face ID)

Go to *Settings › Face ID & Passcode › Change Passcode.*

Choose a *6-digit or alphanumeric code.* Avoid simple combinations like birthdays, "123456," or repeating numbers. Face ID is convenient, but your passcode is the final lock that protects everything.

2. Keep "Stolen Device Protection" On

iOS 18 introduces an enhanced *Stolen Device Protection* feature that prevents thieves from changing critical settings without your Face ID or Touch ID.

Find it under *Settings › Face ID & Passcode › Stolen Device Protection → ON.*

It prevents unauthorized access to passwords, Apple ID sign-outs, and passcode resets—even if someone knows your phone's code.

3. Use "Silence Unknown Callers"

Scammers rely on spam calls to trick users into revealing details or clicking phishing links.

Go to *Settings › Phone › Silence Unknown Callers → ON.*

Calls from your contacts, recent numbers, or Siri suggestions still come through, but strangers go straight to voicemail.

4. Block Scam Texts and Links

Go to *Settings › Messages › Unknown & Spam › Filter Unknown Senders.*

This separates suspicious texts into a different tab and prevents notifications.

If you receive messages claiming to be from Apple or your bank asking for "verification," never tap the links—delete them immediately.

5. Keep Your Software Up to Date

Security fixes arrive quietly with every iOS update. Go to *Settings › General › Software Update* and enable *Automatic Updates.* Many hacks target users running outdated versions.

6. Protect SIM and eSIM Access

Under *Settings › Cellular › SIM PIN,* enable a SIM PIN. This prevents criminals from removing your SIM or transferring your number to another device.

Setting Up Face ID and Passkeys Properly

Face ID is more than a convenience—it's a biometric vault. And with Apple's new *Passkeys*, it becomes the foundation of a passwordless future.

1. Set Up Face ID the Right Way

Go to *Settings › Face ID & Passcode › Set Up Face ID.*
Follow the prompts to scan your face.

- Ensure you're in good lighting.

- Move your head slowly to capture all angles.

- Add an *Alternate Appearance* if you often wear glasses or head coverings.

Face ID can now authenticate:

- App Store and Safari purchases

- Password Autofill

- Banking apps

- Secure Notes and Apple Pay

2. Use Passkeys for Passwordless Logins

Passkeys replace passwords with biometric keys stored securely in iCloud Keychain. They can't be phished or stolen.

To enable:

- Go to *Settings › Passwords.*

- Turn on *Sync with iCloud Keychain.*

- When logging into supported apps or websites, choose *Sign in with Passkey.*

193

Your Face ID becomes your password. No typing, no remembering, no risk of leaks.

3. Lock Down Autofill and Password Access

Under *Settings › Passwords,* toggle *Require Face ID.* This ensures even someone holding your unlocked phone can't see stored credentials without your biometric confirmation.

Two-Factor Authentication for iCloud and Apps

Two-factor authentication (2FA) is the single most powerful defense against account breaches. It ensures that even if someone guesses or steals your password, they can't access your data without physical access to your trusted device.

1. Enable 2FA for Your Apple ID

Go to *Settings › [Your Name] › Password & Security › Two-Factor Authentication → ON.*

Once active, you'll receive verification codes only on trusted devices or phone numbers.

2. Manage Trusted Devices Wisely

In the same menu, check your list of *Trusted Devices.* Remove any you no longer own or use.

This prevents a stolen or sold iPad or Mac from being used to access your codes.

3. Use the Built-In Code Generator for Third-Party Apps

You don't need a separate authenticator app anymore—iOS includes one.

Go to *Settings › Passwords › [App Login] › Set Up Verification Code.*

Scan the QR code provided by the service (like Instagram, Google, or Amazon).

From then on, your iPhone automatically generates fresh codes every 30 seconds.

4. Protect Your Recovery Methods

Add a *Recovery Contact*—someone you trust who can verify your identity if you get locked out of your Apple ID.

Find it under *Settings* › *[Your Name]* › *Sign-In & Security* › *Account Recovery.*

How to Locate a Lost iPhone Even If Powered Off

Losing your iPhone is stressful—but Apple's *Find My network* makes it recoverable even if it's turned off or offline.

1. Enable Find My iPhone

Go to *Settings* › *[Your Name]* › *Find My* › *Find My iPhone* → *ON.* Also enable *Find My Network* and *Send Last Location.*

This lets nearby Apple devices anonymously relay your phone's location back to you—even when it's powered down.

2. How to Track a Lost Device

- Go to iCloud.com/find or open the *Find My* app on another Apple device.

- Tap *Devices* and select your iPhone.

  You'll see its last known location on a map.

If the phone is offline, it will update its position as soon as it reconnects or is powered back on.

3. Mark It as Lost

Tap *Mark as Lost.* This locks your device remotely with a custom message and phone number on the Lock Screen.

Even if someone finds it, they can't access your data or reset your iPhone without your Apple ID credentials.

4. Erase It Remotely (If Recovery Seems Impossible)

If your phone is unlikely to be recovered, choose *Erase This Device.* Your personal data will be wiped clean the next time the device connects to the internet—but the phone will remain locked to your Apple ID.

This feature, called *Activation Lock,* renders the device useless to thieves.

197

Safe Data-Wipe Procedure Before Resale

When it's time to upgrade or sell your iPhone, never just delete photos and call it a day. Even "factory resets" can leave traces if not done properly. Follow this exact process to protect your identity.

1. Back Up Your Data First

- Go to *Settings* › *[Your Name]* › *iCloud* › *iCloud Backup* › *Back Up Now,* or
- Connect to your computer and use Finder to back up locally.

2. Sign Out of iCloud and Services

Go to *Settings* › *[Your Name]* › *Sign Out.*

Enter your password when prompted. This disables *Find My iPhone* and removes Activation Lock.

3. Unpair Devices

If connected, unpair your Apple Watch and AirPods. They retain ownership links until manually removed.

4. Erase Securely

Go to *Settings › General › Transfer or Reset iPhone › Erase All Content and Settings.*

Confirm when prompted.

This fully wipes all data—including Apple Pay cards, biometric data, and saved passwords—then resets the phone to factory settings.

5. Verify the Wipe

After reset, the phone should display the "Hello" setup screen. If it asks for an Apple ID during setup, the iCloud account wasn't properly removed—repeat the process.

6. Physically Clean and Inspect Before Sale

Clean your device gently with a microfiber cloth, remove your SIM (if applicable), and ensure no case residue or identifying stickers remain.

Final Thoughts

Security isn't paranoia—it's peace of mind. The iPhone 17 Pro gives you the tools to defend your privacy, but you are the final gatekeeper.

When your passcode is strong, Face ID is trained correctly, 2FA is active, and Find My is running in the background, your iPhone becomes virtually impenetrable. Even if it's lost or stolen, your personal world remains safe.

And when you finally decide to part with it, wiping it clean ensures the next owner gets a fresh start—while your identity stays yours.

Protecting Your Device from Theft and Scam Calls

Activate a strong passcode and Stolen Device Protection

Setting Up Face ID and Passkeys Properly

Enroll securly for passwordless sign-inin

How to Locate a Lost iPhone Even If Powered Off

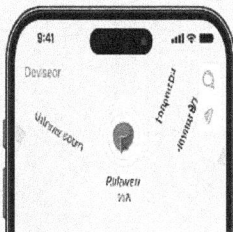

Enable Find My network and mark as lost

Safe Data-Wipe Procedure Before Resale

Erase all content. and settings before

Part IV – Advanced Troubleshooting & Pro Maintenance

Chapter 13

The Complete Troubleshooting Index

Every great device has moments when it needs a little help. The iPhone 17 Pro is no exception. Even with iOS 18's precision engineering, performance hiccups, overheating, or random app behavior can still appear. What matters isn't avoiding these moments—it's knowing how to respond to them calmly and correctly.

This chapter serves as your *quick-access handbook for problem-solving*. Instead of searching the internet or visiting a repair shop for every minor glitch, you'll have the essential knowledge at your fingertips: what to check, what to reset, when to back up, and when to contact Apple Support.

Think of it as your *iPhone First Aid Kit*—practical, clear, and ready whenever your device needs a reset, a recharge, or a recovery.

Quick Reference: "If This Happens → Do This"

Below is a simple mental checklist for the most common iPhone 17 Pro problems. Use it like a guidebook, not a manual—quick, decisive actions often fix more than you think.

1. If Your Battery Drains Too Fast →

- Check *Settings › Battery* for high-drain apps.

- Turn on *Low Power Mode* temporarily.

- Disable *Background App Refresh* for unnecessary apps.

- Restart your iPhone.

  If the problem persists after 24 hours, recalibrate by letting your battery drain to 0%, then fully charge to 100% once.

2. If Your iPhone Feels Hot or Overheats →

- Close unused apps and remove the case temporarily.

- Avoid using it while charging or in direct sunlight.

- Lower brightness and stop background syncing.

204

- If filming or gaming, take a 5–10 minute cooling break. If heat remains constant during normal use, update iOS— thermal bugs often resolve through patches.

3. If Your iPhone Freezes or Becomes Unresponsive →

- Perform a *Force Restart*: Press and release *Volume Up*, then *Volume Down*, then hold *Side Button* until the Apple logo appears.
- Once restarted, check *Settings ⟩ General ⟩ Storage*—a nearly full storage can cause lag.
- If freezing continues, reset all settings (not data): *Settings ⟩ General ⟩ Transfer or Reset iPhone ⟩ Reset ⟩ Reset All Settings.*

4. If Wi-Fi or Bluetooth Disconnects Frequently →

- Toggle Airplane Mode ON for 10 seconds, then OFF.
- Forget and rejoin your Wi-Fi network.

- Reset network settings: *Settings › General › Reset › Reset Network Settings.*

- Ensure your router's firmware is up to date.

  If multiple devices fail on the same network, the router—not your iPhone—is the culprit.

5. If Apps Crash or Won't Open →

- Swipe up to close the app completely.

- Check the App Store for updates.

- Restart your iPhone.

- If it persists, delete and reinstall the app.

  If *many* apps crash after an iOS update, perform a clean reboot and wait overnight—background indexing may be temporarily overloading memory.

6. If iCloud Photos or Files Won't Sync →

- Go to *Settings › [Your Name] › iCloud › Photos* and toggle OFF, wait 30 seconds, then ON again.

- Plug into power and Wi-Fi; syncing pauses when low on battery.

- Check iCloud storage space.

- Log out and back in to iCloud if the issue continues.

7. If Messages or FaceTime Fail to Activate →

- Check Wi-Fi or data connection.

- Make sure *Date & Time › Set Automatically* is ON.

- Turn iMessage and FaceTime OFF, then ON again.

- If dual SIM, ensure the correct line has SMS enabled.

8. If Camera or Flashlight Won't Work →

- Force-quit Camera app, then reopen.

- Restart your iPhone.

- If third-party camera apps fail but Apple's works, delete and reinstall those apps.

  If both still fail, a hardware issue could exist—schedule an Apple inspection.

9. If Audio Sounds Distorted or Muffled →

- Clean the speaker grills gently with a dry toothbrush or soft cloth.

- Check if Bluetooth is accidentally connected to another device.

- Go to *Settings › Sounds & Haptics › Headphone Safety* and disable any limits temporarily.

10. If Your iPhone Won't Charge →

- Inspect cable and adapter for damage.

- Try a different outlet or certified Apple charger.

- Clean the charging port carefully with compressed air or a soft brush.

- If still unresponsive, test with wireless MagSafe charging to isolate port damage.

Battery, Heat, Crash, Wi-Fi, and App Issues Summarized

Let's simplify it even further—the "five core fix principles" that apply to almost every technical issue:

1. **Restart Before Anything Else** — It clears memory, resets temporary processes, and resolves 80% of glitches instantly.

2. **Update Regularly** — Bugs live in outdated versions. Keep both apps and iOS current.

3. **Reduce Background Work** — Too many simultaneous syncs, refreshes, and animations can overload even the best processors.

4. **Clear Storage & Cache** — A cluttered phone behaves like an exhausted one. Maintain at least 10GB free space.

5. **Monitor With Awareness** — If a single app always triggers heat, lag, or crashes, it's the problem—not your phone.

If these principles don't work after repeated use, it's time to back up and reset.

When to Back Up, Reset, or Call Apple Support

Not every problem demands a full reset—but some do. Knowing when to stop troubleshooting and start restoring can save hours of frustration.

Back Up Your iPhone When:

- Battery health drops below 85%.

- You're about to install a major iOS update.

- Apps frequently crash or system files behave unpredictably.

- You're planning to reset, repair, or sell your device.

To back up:

- **iCloud:** *Settings › [Your Name] › iCloud › iCloud Backup › Back Up Now.*

- **Mac/PC:** Connect via USB and select *Back Up Now* in Finder or iTunes.

Reset Your iPhone When:

- Freezes, lag, or unresponsive apps persist even after updates.

- Major functions (Wi-Fi, Bluetooth, Camera) stop working system-wide.

- You've tried every other fix and performance is still poor.

Go to *Settings › General › Transfer or Reset iPhone › Erase All Content and Settings.*

Then restore from your latest backup.

Call Apple Support When:

- You suspect a *hardware defect* (battery swelling, camera failure, port issues).

- You see *unexplained shutdowns* or your phone won't turn on.

- iCloud lock or activation errors persist after sign-in.

- Your warranty or AppleCare coverage might apply.

You can contact Apple Support through the *Support app* or support.apple.com, where chat and call options are available 24/7.

Documenting Your Issue for Warranty Claims

If your iPhone 17 Pro is still under warranty or AppleCare+, documenting your issue clearly can help you receive faster service or free repair. Apple's diagnostic process relies on detailed, reproducible information.

1. Keep a Record of Symptoms

Write down when the issue started, how often it occurs, and what triggers it (e.g., "phone heats up after 10 minutes of video recording").

2. Take Screenshots or Videos

If your screen flickers, your camera fails to focus, or Wi-Fi drops repeatedly, record it. Evidence helps Apple technicians replicate the issue quickly.

3. Note Your iOS Version and Apps Involved

Go to *Settings › General › About* to note your iOS version and affected apps. Include this in your claim or appointment notes.

4. Run Apple Diagnostics (Optional)

Type *diagnostics*:// into Safari (on your iPhone) and follow the on-screen steps. It generates a report code for Apple's remote support team.

5. Bring Your Proof of Purchase or AppleCare+ Plan

If your iPhone was purchased from a third party, ensure your receipt or AppleCare coverage is linked to your Apple ID.

6. Schedule Smartly

Use the *Apple Support App › Get Support › Repairs & Physical Damage* to book an appointment. Having data and photos ready will cut your appointment time in half.

Final Thoughts

Troubleshooting isn't just about fixing problems—it's about understanding your device's language. The iPhone 17 Pro gives subtle warnings before major issues: warmth before shutdowns, lag before crashes, sync pauses before data errors.

By paying attention early and using this chapter as your guide, you can keep your iPhone running smoothly for years without panic or guesswork.

And if a true failure ever occurs, you'll be prepared—with backups, records, and calm confidence that your data, warranty, and workflow are all protected.

If This Happens → Do This

Battery drains too fast
- Check Battery ore-fats Avoid Low Power Mode

IPhone feels hot
- Close apps, Avoid sunlight Take a break

IPhone freezes
- Force-restart. Restarten Resetarten, Settings

App won't open
- Toggle Airplanke mode, Forget the Network, Resel Network Settings

App reun
- Swipe up, Check App Store for Updates, Reinstall

Restart

Update

Reduce Background Work

Clear Storage & Cache

Monitor With Awareness

When to Back Up, Reset, or Call Apple Support

Back Up
- Battery health below 85%, installing major update
- Apps crashing stopp-up run system-wide
- Trying every other fix

Call Apple Support
- Uardware defect
- Unexplated shutdowns or persisting after sin-in
- Warranty or AppleCare or AppleCare coverage

Documenting Your Issue for Warranty Claims

- Keep a record of symptoms
- Take screenshots or videos
- Note iOS version and apps involved
- Run Apple Diagnostics

- Knoιε IOS 'νɔ by diagnostics:// in Satari

Chapter 14

Maintaining Peak Performance Over Time

Your iPhone 17 Pro is designed for longevity — not just to survive, but to perform brilliantly for years. But peak performance isn't automatic. Like a well-kept car, your iPhone thrives when maintained regularly, tuned thoughtfully, and updated with care.

Most people only think about maintenance when something breaks. But consistent, preventive care keeps your phone smooth, battery-strong, and ready for the future. In this final chapter, you'll learn how to create easy weekly and monthly rituals, clear invisible clutter, monitor diagnostics, interpret software updates wisely, and future-proof your data before any major system change.

These habits will help your iPhone feel fast, secure, and stable— long after newer models hit the market.

Weekly and Monthly Maintenance Rituals

A little upkeep goes a long way. You don't need to overhaul your system every week — just keep it clean, organized, and balanced.

Every Week: The 10-Minute Tune-Up

1. **Restart Your iPhone** – A simple restart clears temporary memory (RAM) and background processes that accumulate over days. It's like giving your phone a full night's rest.

2. **Close Idle Tabs and Apps** – In Safari, tap the tabs icon and close unneeded pages. In multitasking view, swipe away apps you haven't used in days.

3. **Review Storage** – Go to *Settings › General › iPhone Storage.* Delete temporary downloads or duplicate photos.

4. **Check for App Updates** – Outdated apps often cause lag or crashes. Visit the App Store and update all apps at once.

5. **Scan Battery Usage** – In **Settings › Battery**, review which apps consumed the most power this week. If one app spikes unusually, check for updates or reinstall it.

216

Every Month: The Deep Clean Routine

1. **Clear Safari Cache** – Go to *Settings › Safari › Clear History and Website Data.* This removes browsing data and frees hidden space.

2. **Offload Unused Apps** – In **Settings › iPhone Storage**, enable *Offload Unused Apps.* It deletes unused apps but preserves your data.

3. **Organize Home Screens** – Declutter icons, review widgets, and remove anything you haven't used in 30 days. Simplicity improves both performance and mental focus.

4. **Recalibrate Battery (Optional)** – Let the battery drop to 5%, then charge uninterrupted to 100%. This helps the system re-learn your battery capacity for more accurate readings.

5. **Back Up Your Data** – Always back up once a month, even if everything seems fine. Backups are your safety net against unexpected issues or future iOS updates.

Quarterly (Every 3 Months): The Stability Check

- Check *Settings › General › About › iOS Version* and ensure your phone is running the latest stable release.

- Clean your charging port and speaker grills gently with a soft brush.

- Run a full diagnostic using *Settings › Battery › Battery Health & Charging.*

- Evaluate performance — if apps take longer to open or the phone feels sluggish, consider resetting settings or clearing excess storage.

Clearing Caches, Checking System Diagnostics, Testing Battery Health

Your iPhone runs thousands of small background operations daily. Over time, these create caches — invisible bits of stored data that can slow things down if left unchecked.

1. Clearing App Caches Safely

Apple doesn't offer a one-button cache cleaner (for security reasons), but you can clear them manually:

- Delete and reinstall large apps (like Instagram or TikTok) periodically. It clears temporary data while preserving your account.

- For Safari: *Settings › Safari › Clear History and Website Data.*

- For Apple Music or Photos: Sync to iCloud and re-download only what you use.

Avoid third-party "cleaner" apps. They often do more harm than good by interfering with Apple's system management.

2. Checking System Diagnostics

If your phone feels slower, it's worth checking its internal diagnostics.

- Go to **Settings › Battery › Battery Health & Charging.**

219

- o *Maximum Capacity* shows your battery's relative health (ideally above 85%).

- o *Peak Performance Capability* should say "Normal."

- To test deeper system health, use *Settings › Privacy & Security › Analytics & Improvements › Analytics Data.* Search for recent "JetsamEvent" files — if you see many, it means your system is running low on memory frequently, suggesting it's time to free up space or restart more often.

3. Testing Hardware Components (Hidden Diagnostics)

Apple's built-in *Diagnostics Mode* can reveal issues you might not notice.

Type *diagnostics*:// in Safari and follow the prompt—it will open Apple's remote testing service. This helps Apple Support quickly identify problems with battery, sensors, or wireless components if needed.

4. When to Consider a Battery Replacement

If your *Battery Health* falls below 80% and your iPhone drains rapidly or shuts down early, consider a replacement. Visit an Apple Authorized Service Provider or Genius Bar; genuine replacements restore both performance and reliability.

Updating Responsibly (How to Read Apple's Update Notes)

Updating your iPhone keeps it secure and stable, but doing it without awareness can backfire. Some users update too quickly—only to face app crashes or reduced battery life. Updating responsibly means understanding what's changing and when to install.

1. Read the Update Notes First

Go to *Settings* › *General* › *Software Update.* Before tapping "Download and Install," tap *Learn More.*

Apple lists:

- *New Features* (like redesigned Control Center or added camera tools)
- *Bug Fixes and Performance Improvements*

- *Security Updates* (critical to privacy)

Pay attention to updates labeled *Security Response* — these are urgent patches for known vulnerabilities and should always be installed immediately.

2. Wait 48 Hours for Major Updates

When a big iOS version drops (e.g., iOS 19), give it two days. This lets Apple fix initial bugs and lets you confirm compatibility through forums or Apple's Support page.

3. Backup Before Every Update

Never install a major update without a full backup.

- For iCloud: *Settings › [Your Name] › iCloud › iCloud Backup › Back Up Now.*
- For Finder: Connect your iPhone to your Mac, open Finder, and choose *Back Up Now.*

4. Plug In and Connect to Wi-Fi

Updating over low battery or cellular data increases the risk of interruptions and corruption. Always connect to a stable Wi-Fi network and power source.

5. Avoid Beta Software Unless You're Testing

Public betas can offer exciting previews—but also carry bugs, app crashes, and higher battery drain. Stick to official releases unless you're comfortable troubleshooting issues.

Preparing for Future iOS Versions Without Data Loss

Apple releases new iOS versions yearly, and preparing properly ensures a smooth transition without losing precious files or memories.

1. Perform a Full Backup Before Major Updates

Before upgrading to a new iOS version, ensure all your data is backed up to both *iCloud* and your *computer*. This dual-layer protection prevents loss even if something fails during installation.

2. Check App Compatibility

Older apps may not be updated for the latest iOS.

Before upgrading:

- Visit the App Store → Check "Version History" to confirm recent updates.

- Remove abandoned apps that haven't been updated in over a year.

3. Free Up Space

Major updates require at least *8–10GB* of free storage.

Go to *Settings › iPhone Storage* and delete unnecessary downloads or temporary files beforehand.

4. Sync Your Media and Documents

Ensure *iCloud Photos*, *Contacts*, and Notes are up to date. Manually open these apps while on Wi-Fi to trigger sync before upgrading.

5. Document Your Layout and Preferences

Take screenshots of your Home Screen layout, widget setup, and

Shortcuts automations. If an update resets your layout (rare but possible), you can rebuild it quickly.

6. Avoid Updating on Day One of Release

Unless a new iOS includes critical security patches, wait a few days. Apple often releases follow-up updates (e.g., iOS 19.0.1) to fix early issues.

7. Verify Backup Integrity After Update

After installation, check *Settings › [Your Name] › iCloud › iCloud Backup › Last Backup Date* to ensure your backups are still functioning.

The Mindset of a Long-Term User

The secret to owning a high-performing iPhone isn't replacing it—it's respecting it. Your device rewards consistency: charge intelligently, restart regularly, and update mindfully. Avoid shortcuts that promise "speed boosts" or "cleaners." Apple's ecosystem thrives on balance and trust, not forced optimization.

A well-maintained iPhone 17 Pro can easily last 5–7 years without feeling obsolete. Its value isn't just in the specs—it's in how well it continues to serve *you*.

Every restart, every backup, every moment of attention is a quiet investment in that longevity.

So take five minutes each week to care for it—and it will return years of flawless performance, beautiful memories, and peace of mind.

Your iPhone will stay what it's always been: not just a piece of technology, but a reflection of the way you live—organized, secure, and endlessly capable.

Weekly and M Maintenance

- Restart the iPhone
- Review storage and battery usage
- Update apps

Clearing Caches, Checking Diagnostics

Battery Health

Maximum Capacity 95%

- Clear Safari history
- Offload unused apps
- Run diagnostics

Updating Responsibly

Settings Software Update

105 Update

Learn More

- Read update notes
- Wait to install new iOS versions
- Back up before updating

Preparing for Future iOS Versions

- Perform a full backup
- Sync media and documents
- Ensure app compatibility

Chapter 15

Repair, Warranty & Apple Support Essentials

Even the most reliable devices can need help one day. Whether it's a cracked screen, battery degradation, or a mysterious glitch, knowing how to navigate Apple's repair system can save you time, money, and stress.

The iPhone 17 Pro is built with exceptional durability and intelligent self-diagnostics, but understanding the support ecosystem around it—the warranties, service options, and consumer protections—empowers you to get the right help without confusion.

This chapter is your complete handbook for when things go wrong: how Apple's warranty system really works, how to prepare for a repair, how to use the Apple Support app for real-time solutions, how to share diagnostic logs with technicians, and what rights you have as a customer—no matter where you live.

Understanding Warranty Tiers and AppleCare Options

When you buy an iPhone, you automatically receive a **standard Apple Limited Warranty**. It covers manufacturing defects and hardware failures for **one year** from the date of purchase, along with **90 days of complimentary technical support**. But that's only the foundation—AppleCare+ and regional protections extend this significantly.

1. Apple Limited Warranty (Standard)

- Duration: 1 year (hardware coverage), 90 days (technical support).

- Covers: Manufacturing defects, non-accidental hardware faults, battery performance issues that fall below expected capacity (less than 80% during warranty).

- Does *not* cover: Accidental damage, liquid exposure, cosmetic issues, or software corruption.

2. AppleCare+ (Extended Coverage)

AppleCare+ upgrades your warranty to *two years* and includes *accidental damage coverage* (for a small service fee per incident). It's essentially your iPhone's insurance plan through Apple.

Benefits:

- Priority access to Apple Support.
- Hardware protection for up to two accidental incidents per year.
- Express Replacement Service (Apple ships a replacement before you return your damaged device).
- Extended battery service (if capacity drops below 80%).

3. AppleCare+ with Theft and Loss (Select Regions)

Available in the U.S., UK, Canada, and select markets, this plan adds coverage for *theft or loss.* If your iPhone is stolen or unrecoverable, Apple will replace it for a small deductible fee, provided *Find My iPhone* was enabled at the time of loss.

4. Out-of-Warranty Repairs

If your warranty has expired or the damage isn't covered, Apple still offers flat-rate repairs through Apple Authorized Service Providers (AASPs). These use genuine Apple parts and include a 90-day service guarantee.

Tip: Always check your coverage before scheduling a repair.

Go to *Settings › General › About › Coverage,* or visit checkcoverage.apple.com and enter your device serial number.

How to Prepare Before Visiting a Repair Center

Walking into a repair center unprepared often means delays. With a little preparation, you can make your visit fast and productive—and ensure that your data remains safe.

1. Back Up Your iPhone

Before handing your device to anyone, always back it up:

231

- **iCloud:** *Settings › [Your Name] › iCloud › iCloud Backup › Back Up Now.*

- **Mac or PC:** Connect your phone and back up via Finder or iTunes.

  This ensures you can restore everything afterward, even if your iPhone is replaced.

2. Disable Find My iPhone

Apple technicians can't service a locked device.

Go to Settings › [Your Name] › Find My › Find My iPhone → OFF.

You'll need to enter your Apple ID password to confirm.

3. Gather Necessary Items

Bring:

- Your iPhone and original receipt or proof of purchase.

- AppleCare+ or insurance paperwork (if applicable).

- A valid photo ID.

- Any accessories relevant to the issue (like a charger if charging problems persist).

4. Reproduce the Problem Beforehand

Try to recreate the issue so you can describe it clearly. Note how often it happens, what triggers it, and whether it occurs with specific apps or conditions. The more specific you are, the faster technicians can diagnose it.

5. Clean and Power Down

Before visiting, gently clean your phone with a microfiber cloth and turn it off. This not only helps staff but also prevents accidental inputs during diagnostics.

Using the Apple Support App for Live Troubleshooting

You don't always need to visit a repair center—many problems can be solved remotely through the *Apple Support app.* It's a free, built-in help desk that connects you directly with Apple's specialists via chat, phone, or screen-sharing.

1. Download or Open the App

Find it preinstalled or download it from the App Store. Sign in with your Apple ID to automatically display all your registered devices.

2. Navigate to Your Device

Tap *My Devices › iPhone 17 Pro.* You'll see sections like *Repairs & Physical Damage*, *Battery & Charging*, *Performance*, *Apps & Software,* and *Connectivity.*

3. Run Basic Troubleshooting

The app walks you through step-by-step instructions for each issue. Many fixes—like resetting network settings, clearing caches, or updating firmware—can be done right there.

4. Chat or Call Apple Experts

If the guided help doesn't solve your problem, tap *Chat with an Expert* or *Call Apple Support.* You'll connect to a specialist who can remotely diagnose your device.

5. Schedule Repairs or Mail-In Service

From the same interface, you can schedule an in-store appointment or arrange a *mail-in repair* where Apple sends you a prepaid box for your iPhone.

6. Track Repair Progress

Once your repair begins, you can monitor progress directly in the Support app under *My Repairs*. You'll see the current status (e.g., *Received*, *In Service*, *Ready for Pickup*).

Exporting Diagnostics Logs to Speed Up Help

When troubleshooting complex issues (like battery drain or hardware sensor failures), diagnostics logs give Apple Support a detailed snapshot of your phone's internal behavior. Sharing these logs helps engineers pinpoint the issue faster.

Here's how to do it safely:

1. Generate the Logs Automatically

Go to *Settings › Privacy & Security › Analytics & Improvements.*

Enable *Share iPhone Analytics.*

This allows Apple to collect anonymous logs, but you can also export them manually.

2. Access Diagnostic Reports

- Open *Settings › Privacy & Security › Analytics Data.*
- Scroll through the list to find files labeled "log-aggregated" or "diagnostic." These contain crash reports, thermal data, and background activity logs.

3. Export the Logs

Tap the file → *Share* → choose *Mail* or *Messages.* Send it to Apple Support if requested. You can also AirDrop it to your Mac for personal review.

4. Run a Live Diagnostic Session

During a support call, Apple can initiate remote diagnostics. You'll receive a link via text or email; tapping it runs a quick scan and sends encrypted results back to the technician. This can detect failing batteries, damaged sensors, or software conflicts instantly.

Your Consumer Rights (For Different Regions)

While Apple offers a global warranty framework, your rights can vary based on where you live. Here's a summary of what consumers are entitled to in major regions:

United States

- Federal law offers limited consumer protection, but Apple complies with local state laws (like California's extended warranty obligations).
- AppleCare+ provides nationwide coverage for two years with accidental damage protection.

European Union & UK

- EU and UK law guarantees a *minimum two-year warranty,* regardless of AppleCare status. This is *in addition* to Apple's standard warranty.

- Faults present at purchase or emerging within the first two years are Apple's responsibility to fix or replace.

Canada

- Provincial consumer laws ensure "fit for purpose" protection beyond the manufacturer's one-year warranty, depending on the province.

Australia & New Zealand

- Under Australian Consumer Law, Apple must repair, replace, or refund faulty products even after the standard warranty expires if a "reasonable consumer" wouldn't expect the failure so soon.

Asia (e.g., Singapore, Japan, UAE, Nigeria)

- Warranty terms match Apple's one-year standard, with AppleCare+ available in select regions.

- Local consumer laws vary, but most follow similar protection principles against manufacturing defects and early hardware failures.

Tip:

Always retain your purchase invoice, even for devices bought online or abroad. Apple's coverage is global, but proof of purchase helps link your serial number to your region's laws.

Final Thoughts

Apple's repair ecosystem is one of the most transparent in the tech world—if you know how to use it. The real secret to smooth support isn't just coverage—it's preparation. Back up your data, know your rights, use the Support app, and provide clear logs.

Whether your iPhone 17 Pro needs a screen replacement or a subtle software recalibration, you'll move through the process confidently, without panic or confusion.

Because the truth is, a well-informed user doesn't just own their device—they understand it. And with that understanding comes the ultimate power: control over your technology, your privacy, and your peace of mind.

Repair, Warranty & Apple Support Essentials

Understanding Warranty Tiers and AppleCare Options

Apple Limited Warranty   AppleCare+

How to Prepare Before Visiting a Repair Center

Back Up   Turn Off "Find My   Clean

Using the Apple Support App for Live Troubleshooting

Chat or Call for Assist Schedule Repiar

Exporting Diagnostics Logs to Speed Up Help

Share Analytics Data   Run Diagnostics

Your Consumer Rights (For Different Regions)

Consumer Protection Laws By Region
But Universally Cover Defects

Part V – Beyond the Basics

(Longevity, productivity, and creativity.)

Chapter 16

Smart Integrations

One of the most powerful but often overlooked aspects of owning an iPhone 17 Pro is how effortlessly it connects with everything else around it. Apple's secret sauce isn't just hardware or software — it's the *harmony* between devices. The magic happens when your iPhone, iPad, Mac, and Apple Watch all speak the same fluent language of convenience and continuity.

This chapter is all about that harmony. You'll learn how to use Continuity, AirDrop, Handoff, and Universal Clipboard to move seamlessly across devices; how to pair your iPhone 17 Pro with an Apple Watch, Mac, or iPad for a synchronized life; and how third-party apps can supercharge your daily workflow without breaking Apple's elegant ecosystem.

When you finish this chapter, you'll understand not just how to use your iPhone — but how to let it *work with everything else you own.*

Using Continuity, AirDrop, Handoff, and Universal Clipboard

Apple's *Continuity suite* is the invisible thread that connects your digital life. Once enabled, it allows your iPhone, iPad, and Mac to share tasks instantly — like picking up a note on one device right where you left off on another.

1. Continuity in Everyday Life

Continuity isn't a single feature — it's a collection of small bridges that make big differences. Imagine writing an email on your Mac and finishing it on your iPhone without saving drafts, or copying text on your iPhone and pasting it instantly on your Mac. That's Continuity in action.

To make it work:

- All your devices must be signed in to the *same Apple ID.*

- Keep *Wi-Fi and Bluetooth ON* on all devices.

- Ensure devices are near each other (within 10 meters).

2. AirDrop — The Fastest File Sharing System on Earth

Forget email attachments and messaging apps — *AirDrop* lets you send photos, documents, or videos instantly between Apple devices using peer-to-peer Wi-Fi.

To Use AirDrop:

- Open *Control Center → long-press the top-left network box → AirDrop → Everyone (or Contacts Only).*

- Select the file you want to share, tap the *Share* icon, and choose your nearby device.

- On the receiving device, tap *Accept.*

AirDrop is especially powerful for:

- Sharing large videos without compression.

- Moving entire folders between devices.

- Sending documents securely (no cloud middleman).

245

Pro Tip: For privacy, switch AirDrop back to *Contacts Only* after sharing — it prevents unsolicited requests from strangers in public places.

3. Handoff — The Seamless Task Transfer

Handoff lets you begin a task on one Apple device and instantly continue it on another — exactly where you left off.

Example: You're reading an article in Safari on your iPhone, then open your MacBook — Safari's icon appears on your Dock. One click, and the page appears.

To Enable Handoff:

- On iPhone: *Settings › General › AirPlay & Handoff › Handoff → ON.*

- On Mac: *System Settings › General › AirDrop & Handoff → Allow Handoff.*

Handoff works with:

- Mail, Safari, Notes, Calendar, and Messages.

- Apple Maps (plan routes on one device, continue on another).

- Third-party apps like Evernote, Todoist, and Microsoft Word (if developers support it).

Pro Tip: Handoff also lets you transfer calls between devices. Start a call on your iPhone, and tap the banner on your Mac or iPad to continue hands-free.

4. Universal Clipboard — Copy Once, Paste Anywhere

This feature feels like telepathy for your Apple devices.

Copy a paragraph on your iPhone — then press *Cmd + V* on your Mac, and it appears instantly.

How to Enable:

Ensure *Wi-Fi, Bluetooth, and Handoff* are all active on every device.

Then just copy as usual (long press → Copy on iPhone) and paste (Cmd + V or long press → Paste) on another.

Supported Content:

- Text, images, links, phone numbers, and files.

- Even formatted text and entire photos from apps like Notes or Safari.

Pro Tip: Universal Clipboard times out after 120 seconds — so paste quickly!

Pairing with Apple Watch, Mac, or iPad

Your iPhone 17 Pro becomes exponentially more powerful when paired with other Apple devices. Each connection unlocks new levels of automation, wellness, and productivity.

1. Apple Watch — Health, Alerts, and Unlocking Magic

Pairing your Apple Watch transforms your iPhone into a health hub and command center.

To Pair:

- Turn on your Apple Watch and bring it close to your iPhone.

248

- Tap *Continue* when the pairing screen appears.

- Follow on-screen instructions or use the Watch app manually.

Benefits of Pairing:

- **Unlock with iPhone:** Your iPhone can automatically unlock your Apple Watch (and vice versa).

- **Health Sync:** All fitness and health data flows to your iPhone's *Health* app — heart rate, workouts, sleep, and even ECG readings.

- **Ping to Locate Devices:** Swipe up on your Watch's Control Center and tap the iPhone icon — your iPhone will ring even in silent mode.

- **Camera Remote:** Use your Apple Watch to preview and trigger photos remotely on your iPhone 17 Pro's camera.

Pro Tip: Enable *Unlock with Apple Watch* under *Settings › Face ID & Passcode* — it allows your Watch to unlock your iPhone while wearing a mask or sunglasses.

2. Mac — A Perfect Extension of Your iPhone

Your iPhone and Mac form a powerful creative and professional duo.

Key Integrations:

- **Calls and Texts on Mac:**

  Enable *Settings › FaceTime › Calls from iPhone* and *Messages › Text Message Forwarding.*

  You can then make calls or reply to texts right from your Mac, even if your iPhone is charging in another room.

- **Instant Hotspot:**

  No need to enter passwords. Open your Mac's Wi-Fi menu and click your iPhone's name under *Personal Hotspot* — it connects instantly.

- **iPhone as Camera (Continuity Camera):**

  Your iPhone 17 Pro doubles as a high-quality webcam. Mount it on your Mac's display, and open FaceTime, Zoom, or QuickTime — it automatically connects for stunning video clarity.

- **File Sharing and Universal Control:**

  With *Universal Control,* drag your mouse from your Mac's screen straight onto your iPad or even your iPhone (connected via Sidecar).

  Copy text or images and paste across devices as if they were one workspace.

3. iPad — The Creative and Productivity Partner

The iPad pairs perfectly with your iPhone for notes, design, and entertainment.

Core Benefits:

- **Notes Sync:** Start a note on your iPhone and draw or annotate on your iPad using the Apple Pencil — the content syncs instantly.

- **iCloud Tabs in Safari:** See open browser tabs from your iPhone on your iPad, and continue browsing.

251

- **Universal Clipboard:** Copy an address or snippet on iPhone, paste it into iPad Maps or Mail instantly.

- **Sidecar (Mac + iPad + iPhone Trio):** Extend your Mac workspace to your iPad while using your iPhone as a portable command tool.

Pro Tip: Use your iPad as a second display for your Mac, and your iPhone as the wireless remote or document scanner for your workflow.

Third-Party App Ecosystems That Enhance Workflow

Apple's native tools are powerful, but third-party developers expand the iPhone 17 Pro's ecosystem into a true productivity powerhouse.

1. Cloud Storage and Collaboration

- **Notion & Evernote:** Sync notes, projects, and ideas across all devices.

- **Google Drive & Dropbox:** Universal file access for mixed-device teams.

- **Microsoft 365:** Perfect integration for professionals using Macs and PCs.

2. Automation and Workflow Management

- **IFTTT (If This Then That):** Automate repetitive tasks — e.g., "If I arrive at the office, silence my phone."
- **Zapier:** Connect Apple Reminders, Gmail, and Trello for business automation.
- **Shortcuts + Third-Party Apps:** Combine native iOS Shortcuts with Spotify, Slack, or ChatGPT to create custom voice-activated workflows.

3. Creative Ecosystem Enhancers

- **Adobe Creative Cloud:** Sync photos from iPhone directly to Lightroom or Photoshop on Mac.
- **LumaFusion:** Edit 4K video on iPhone, then continue on iPad.

- **Procreate Pocket:** Draw sketches that instantly sync to iPad for detailed illustration.

4. Smart Home and Lifestyle Integration

- **HomeKit & SmartThings:** Control lights, fans, or cameras via Siri.
- **CarPlay:** Connect your iPhone to your car's display for safer, smarter navigation.
- **Find My Network:** Track AirTags, devices, or even your family members' locations securely in real time.

5. AI and Voice Tools

- **ChatGPT App & Siri Shortcuts:** Pair AI assistance with Siri's voice triggers for personalized responses.
- **Otter.ai & Voice Record Pro:** Record meetings or lectures, transcribe instantly, and sync across devices.

Pro Tip: The best integrations are invisible. Automate once, then forget about them — your devices will take care of the rest.

Final Thoughts

The true genius of Apple's design isn't the iPhone itself—it's the *ecosystem* it lives in. When your devices talk to each other, you work faster, move easier, and live more connected without even thinking about it.

From transferring a photo in seconds to answering a call on your Mac, or using your Watch to unlock your phone, these integrations quietly eliminate friction. They make your technology disappear into your daily rhythm — exactly as Apple intended.

The iPhone 17 Pro, when fully integrated, isn't just a phone. It's your personal network: intelligent, responsive, and always in sync with your life.

Using Continuity, AirDrop, Handoff, and Universal Clipboard

Pairing with Apple Watch, Mac, or iPad

Third-Party App Ecosystems that Enhance Workflow

Chapter 17

For Seniors and New Users

The iPhone 17 Pro is more than a high-end device — it's a companion for independence, connection, and ease of life. But for many seniors or first-time users, the transition to such advanced technology can feel intimidating. Apple anticipated this, designing accessibility and control features that make every task — from answering a call to reading a message — as simple and natural as possible.

This chapter is your complete comfort guide. You'll learn simplified gestures, discover how Voice Control and Siri Shortcuts can replace complex actions, and master accessibility shortcuts that make daily use effortless. You'll also learn how to manage screen time, adjust call volume, and set up emergency contacts — so your iPhone becomes not just a device, but a trusted helper.

Simplified Gestures

The iPhone 17 Pro eliminates physical buttons for most actions, relying on intuitive gestures instead. Once you know the basics, navigation becomes second nature.

1. Home and App Switching Made Simple

- **Go Home:** Swipe *up from the bottom edge* of the screen.

- **View Recent Apps:** Swipe up *and pause halfway.* You'll see open apps — swipe left or right to switch.

- **Force Close an App:** In the Recent Apps view, swipe the app *up and off the screen.*

2. Notifications and Control Center

- **View Notifications:** Swipe *down from the top-left corner.*

- **Open Control Center:** Swipe *down from the top-right corner.*

  Here, you can adjust brightness, volume, Wi-Fi, and Bluetooth without opening Settings.

3. Access Search and Widgets

- **Search Anything:** Swipe *down from the middle of the Home Screen* to open the search bar.

- **View Widgets:** Swipe *right from the first Home Screen* for quick access to calendar, weather, or reminders.

4. Screen Zoom and Magnifier

If small text is difficult to read, activate zoom:

- Go to *Settings › Accessibility › Zoom → ON.*

- Double-tap with *three fingers* to zoom in or out.

  You can move around the screen by dragging with *three fingers.*

Pro Tip: Enable *Display Zoom* under *Settings › Display & Brightness › Display Zoom › Larger Text.*

It enlarges icons and menus without distorting the interface.

Voice Control, Siri Shortcuts, and Accessibility Shortcuts

For seniors and beginners, the iPhone's greatest feature is its ability to listen, understand, and act — even without touch.

1. Voice Control — Hands-Free Navigation

Voice Control allows you to use your iPhone entirely with your voice.

Go to *Settings › Accessibility › Voice Control → ON.*

You'll see a blue microphone icon appear — now you can say commands like:

- "Go Home."
- "Open Photos."
- "Swipe left."
- "Scroll down."
- "Tap Share."

You can even dictate text messages or notes without touching the keyboard.

Pro Tip: Create custom commands, such as "Good morning" to open Weather, Calendar, and Messages all at once.

2. Siri Shortcuts — Your Personal Assistant Simplified

Siri isn't just for weather or directions — she can automate complex tasks into single phrases.

Examples:

- "Hey Siri, call my daughter."

- "Hey Siri, send a message to Mark."

- "Hey Siri, read my new messages."

- "Hey Siri, open YouTube."

- "Hey Siri, take me home."

To create your own:

- Go to *Settings › Siri & Search › My Shortcuts › + Add Shortcut.*

- Choose from suggested automations or record your own command.

For daily life:

- "Hey Siri, good night" → turns on Do Not Disturb, lowers brightness, and locks the screen.

- "Hey Siri, medication time" → opens your health reminders app.

3. Accessibility Shortcuts — *One Tap, Multiple Actions*

The Accessibility Shortcut lets you activate helpful tools instantly using the side button.

To enable:

- Go to *Settings › Accessibility › Accessibility Shortcut.*
- Choose features like:
    - *Magnifier* (turns the camera into a zoom lens)
    - *Reduce Motion* (calms animations)
    - *VoiceOver* (reads on-screen text aloud)
    - *Color Filters* (for better visibility)

Once set, *triple-click the Side Button* to access your chosen tools anytime.

Pro Tip: Combine Accessibility Shortcuts with *Back Tap* (Settings › Accessibility › Touch › Back Tap).

You can double-tap the back of your phone to perform actions like opening the flashlight or taking a screenshot.

Managing Screen Time, Call Volume, and Emergency Contacts

Beyond convenience, your iPhone 17 Pro is built to *protect* you — from eye strain, loud noises, and even emergencies.

1. Managing Screen Time and Focus

Excessive screen exposure can be tiring for older eyes.

- Go to *Settings › Screen Time.*
  - Tap *App Limits* to control how long certain apps stay open.

- o Tap *Downtime* to set quiet hours where only essential apps work.

- Use *Focus Modes* to control notifications. For example:

  - o *Sleep Focus* silences everything except calls from family.

  - o *Personal Focus* hides work apps during rest hours.

Pro Tip: In *Settings › Display & Brightness*, turn on *True Tone* and *Night Shift* to adjust color temperature automatically — reducing glare and blue light strain.

2. Controlling Call Volume and Hearing Safety

- Use the *Side Volume Buttons* to adjust call volume while talking.

- Go to *Settings › Sounds & Haptics* to fine-tune ringtones and vibration strength.

- Enable *Headphone Safety* under *Settings › Sounds & Haptics › Headphone Safety*.

It automatically lowers volume if it exceeds safe listening levels.

If hearing is a concern, try:

- *Settings › Accessibility › Audio/Visual › Headphone Accommodations* — it tailors sound frequencies to your hearing range.
- Enable *Live Listen* to turn your iPhone into a remote microphone for AirPods — helpful in noisy environments.

3. Setting Up Emergency Contacts (Critical for Seniors)

Your iPhone's *Medical ID* and *Emergency SOS* features can save lives by giving first responders immediate access to essential information — even when your phone is locked.

To Set It Up:

- Go to *Health App › Profile Icon › Medical ID › Edit.*
- Add:
  - Emergency contacts (family, doctor, or caregiver).

o   Health conditions (allergies, medications, blood type).

o   Choose *Show When Locked* so rescuers can view it from the Lock Screen.

To Trigger Emergency SOS:

- Hold the **S***ide Button* + *Volume Button* until you see the *SOS slider*.

- Continue holding to call local emergency services automatically.

Your emergency contacts will receive your location once the call ends.

Pro Tip: Pair your iPhone with an Apple Watch for fall detection and auto emergency calls — ideal for seniors living alone.

Final Thoughts

Technology should never feel overwhelming — it should feel like *freedom*. The iPhone 17 Pro, when properly customized, becomes more than a device; it becomes an extension of comfort, safety, and connection.

For seniors and new users, mastering a few gestures, setting up Siri, and enabling accessibility tools transforms the experience from confusing to empowering. It's not about keeping up with technology — it's about making technology *keep up with you*.

The beauty of Apple's design is that it scales gracefully with every user. Whether you're 17 or 77, your iPhone 17 Pro can fit your pace, your lifestyle, and your needs — quietly supporting you every day, without ever asking you to change.

FOR SENIORS AND NEW USERS

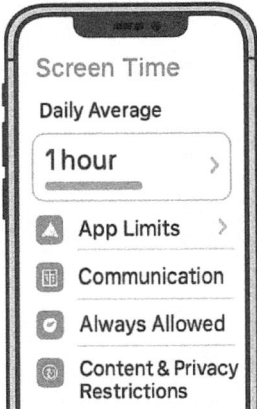

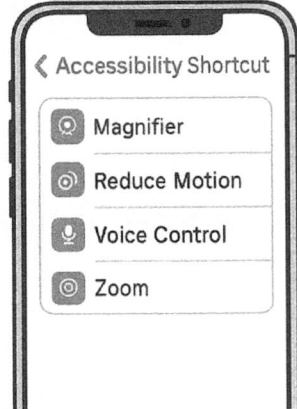

Settings

General

Accessibility >

Zoom

MANAGING SCREEN TIME, CALL VOLUME AND EMERGENCY CONTACTS

VOICE CONTROL, SIRI SHORTCUTS, AND ACCESSIBILITY SHORTCUTS

IOS 17

Turn up thihe volume

.Open Mail

Accessibility >

FOR SENIORS AND NEW USERS

Screen Time

Daily Average

1 hour >

App Limits >

Communication

Always Allowed

Content & Privacy Restrictions

< Accessibility Shortcut

Magnifier

Reduce Motion

Voice Control

Zoom

Chapter 18

For Content Creators & Professionals

The iPhone 17 Pro is no longer just a smartphone — it's a fully capable production studio that fits in your pocket. Whether you're a YouTuber, social media creator, entrepreneur, or digital professional, this device can film cinematic videos, record podcasts, edit professional reels, and publish directly to your platform of choice — all without touching a laptop.

In this chapter, you'll learn how to harness the iPhone 17 Pro's creative power to its fullest. We'll cover how to shoot professional-quality videos, edit seamlessly with mobile apps like iMovie, CapCut, and LumaFusion, and create an efficient cloud-based workflow tailored for YouTubers, influencers, and business owners who want to stay creative on the go.

Shooting Professional-Quality Videos on iPhone 17 Pro

Apple's cinematic ecosystem has matured to the point that filmmakers, vloggers, and educators are producing entire features on iPhones — and with the iPhone 17 Pro's camera hardware, the results are extraordinary. Let's break down how to get studio-level quality in every shot.

1. Master the Camera App — Go Beyond Auto Mode

Apple's Camera app is simple on the surface, but packed with pro features under the hood.

Key Features to Use:

- **ProRes Video** — Enables ultra-high-quality footage ideal for editing.
    - Go to *Settings › Camera › Formats › Apple ProRes → ON*.
    - Use this mode when filming professional or commercial content.

- **4K at 60 fps** — Delivers sharp, smooth visuals for cinematic or YouTube content.

- **Cinematic Mode (for Depth Focus)** — Perfect for storytelling. It simulates a DSLR's bokeh blur and lets you refocus later in editing.

- **Action Mode** — For handheld vlogging or sports footage; reduces shake even while running.

- **Macro and Telephoto Control** — Lets you capture intricate close-ups or stable zooms with minimal distortion.

Pro Tip: Lock your focus and exposure before filming. Tap and hold the screen until "AE/AF Lock" appears. This prevents lighting flickers during recording.

2. Optimize Lighting and Angles

Even the best camera needs good light.

- **Natural Light:** Film near windows or outdoors during the golden hour (early morning or before sunset).

271

- **Artificial Light:** Use a small ring light or LED panel for even tone and clarity.

- **Avoid Harsh Shadows:** Move your light source 45° from your subject's face.

Angles:

- Eye-level shots for tutorials or interviews.

- Slightly elevated shots for vlogs.

- Low-angle shots for dramatic scenes or product showcases.

Pro Tip: Enable *Grid Lines* in *Settings › Camera › Grid* — it helps compose your frame using the rule of thirds for balanced visuals.

3. Audio Makes or Breaks Your Video

Crisp visuals lose their power with bad sound. Always ensure your audio is clean and consistent.

- For speech, use an external lavalier mic or wireless mic system.

- In quiet indoor setups, even wired EarPods with mic produce great clarity.

- For ambient videos or podcasts, record audio separately using the *Voice Memos app* and sync it later in editing.

Pro Tip: To remove background noise live, go to *Control Center ›
Mic Mode › Voice Isolation* before recording.

4. Stabilize and Frame Like a Pro

- Use a *tripod or gimbal* for steady shots.

- For handheld videos, brace your elbows to your body and move smoothly.

- Use *2x zoom* instead of walking closer—it maintains perspective and stabilizes motion.

5. Camera Accessories That Transform Your Workflow

- **Gimbal:** DJI Osmo Mobile or Zhiyun Smooth 5 for cinematic movement.

- **Tripod:** Compact, adjustable stand with Bluetooth remote.

- **Lighting:** Small LED ring light or portable RGB light.

- **Microphone:** Rode Wireless GO II or Shure MV88.

- **Lenses:** Moment or Sandmarc external lenses for wide, macro, or anamorphic effects.

Editing with Photos, iMovie, CapCut, or LumaFusion

You don't need a Mac to create professional content. The iPhone 17 Pro has enough processing power to handle 4K multi-layer video editing — right in your hand. Here's how to choose and use the right app for your creative goals.

1. Photos App (Quick Edits & Social Clips)

Perfect for small touch-ups or trimming reels.

- Open a video → Tap *Edit.*

- Adjust exposure, color, and contrast manually.

- Tap *Filters* for consistent tone or branding (e.g., Warm, Dramatic, Vivid).

- Use *Crop* → *9:16* for vertical videos or *16:9* for YouTube.

  Ideal for: Instagram stories, TikToks, and short reels.

2. iMovie (Beginner-Friendly Storytelling)

Apple's free editor is ideal for beginners who want simple cuts and transitions.

- Import clips → Drag to reorder.
- Add *Themes* for transitions and *Titles* for intros.
- Insert *Voiceovers* or background music directly from your iPhone's library.
- Export in 1080p or 4K.

  Ideal for: *Tutorials, short films, educational videos, and family projects.*

Pro Tip: Pair with *Storyboards* mode to use Apple's built-in templates for vlogs, reviews, and product showcases.

3. CapCut (Social Media Powerhouse)

CapCut is the creator's favorite for TikTok, Instagram, and YouTube Shorts.

- Add clips → Tap *Auto Captions* for subtitles.

- Use *Effects* and *Transitions* for motion sync with music.

- *Keyframe Animation* adds zooms or pans that feel cinematic.

- Built-in templates allow quick trend-based edits.

  Ideal for: *Influencers, vloggers, and short-form content creators.*

Pro Tip: Use *Background Remover* to isolate yourself from messy backgrounds — no green screen needed.

4. LumaFusion (Professional Mobile Editing Suite)

LumaFusion is the iPhone's Final Cut Pro equivalent — used by journalists, filmmakers, and YouTubers.

- Supports multi-track 4K editing.

- Add music, sound effects, titles, LUTs, and overlays.

- Use *Keyframing* for precise motion graphics.

- Sync with *external SSDs* for large project files.

  Ideal for: *Documentaries, commercials, cinematic projects, and long-form YouTube videos.*

Pro Tip: Export your project as an *XML file* to continue editing on Final Cut Pro or DaVinci Resolve later.

Cloud Workflow for YouTubers, Influencers, and Business Owners

The modern creator's greatest strength is mobility. Whether you're uploading from Lagos, London, or Los Angeles, cloud workflows ensure your creative files are always accessible and secure.

1. Cloud Storage You Can Rely On

- **iCloud Drive:** Automatically syncs your photos, videos, and LumaFusion projects. Perfect for solo creators using multiple Apple devices.

- **Google Drive / Dropbox:** Great for collaboration and team access.

- **Adobe Creative Cloud:** If you use Lightroom or Premiere Rush, your edits stay synced across devices automatically.

Pro Tip: Store raw footage in *external SSDs* (like Samsung T7) and keep only active projects in the cloud.

2. Collaborating with Teams or Clients

- Use *Shared Albums* in the Photos app for quick video approvals.

- Create *Shared Folders* in iCloud or Drive for editing handoffs.

- Use *Notion or Trello* to manage content calendars and brand partnerships.

- Record raw clips, upload to the shared folder, and let editors or social managers handle the rest.

3. Uploading to Platforms Efficiently

For creators, time is currency. Streamline your uploads:

- Edit in 4K, but export in *1080p (60fps)* for social media balance between quality and speed.

- Use *YouTube Studio App* for titles, thumbnails, and analytics on the go.

- For Instagram or TikTok, use *CapCut's export presets* to meet platform specs automatically.

4. Protecting Your Work

Never rely on one storage method.

- Keep one *local copy* (external drive), one *cloud copy,* and one *device copy.*

- Enable *iCloud Photos › Optimize iPhone Storage* to save space while preserving access to your full-resolution files online.

Pro Tip: Watermark your branded content subtly before uploading — especially for high-value reels or ads.

Final Thoughts

In the hands of a creator, the iPhone 17 Pro is more than a phone — it's a camera, editor, studio, and distributor in one. Every great idea now has no barrier between imagination and creation.

Whether you're recording your first vlog, editing client footage from a café, or building a brand with global reach, your iPhone empowers you to move at the speed of creativity.

Mastering its tools means mastering freedom — to create anywhere, anytime, at professional quality.

The iPhone isn't just for consuming content. It's for *creating it.*

Shooting Professional-Quality Videos on iPhone 17 Pro

Editing with Photos, iMovie, CapCut, or LumaFusion

Cloud Workflow for YouTubers, Influencers and Business Owners

ProRes Video

Keyframe Animation

Part VI – Appendices

Appendix A: Quick Fix Flowcharts

(Your One-Page Rescue Guide for Common iPhone Issues)

Technology doesn't wait, and neither should you. These quick-fix flowcharts summarize the most common iPhone 17 Pro problems—battery, connectivity, and crashes—into clear, step-by-step decision paths. Each one helps you troubleshoot efficiently before resorting to a full reset or support visit.

Battery Issues Flowchart

Problem: "My iPhone battery drains too fast or won't charge properly."

1. **Step 1: Check Battery Usage**

   → *Settings › Battery › Battery Usage by App*

   → If one app consumes more than 25%, update or reinstall it.

2. **Step 2: Check Charging Setup**

   → Use an original cable and Apple-certified adapter.

   → Try another wall socket or USB port.

3. **Step 3: Is Your Phone Warm While Charging?**

   → Yes → Remove case & keep away from heat.

   → No → Continue.

4. **Step 4: Check Battery Health**

   → *Settings › Battery › Battery Health & Charging*

   → If under 80%, plan for replacement.

5. **Step 5: Reset & Update**

   → Restart phone.

   → Update to latest iOS version (many updates improve battery stability).

Still Draining Fast?

→ Turn on *Low Power Mode* and disable *Background App Refresh.*

→ If unchanged, contact Apple Support for diagnostic testing.

Connectivity Issues Flowchart

Problem: "Wi-Fi, Bluetooth, or mobile data keeps dropping."

1. **Step 1: Check Airplane Mode**

   → If ON, turn it OFF. Wait 10 seconds.

2. **Step 2: Reset Network Settings**

   → *Settings › General › Reset › Reset Network Settings*

   → Reconnect to your Wi-Fi and re-enter password.

3. **Step 3: Check APN Settings (Mobile Data)**

   → *Settings › Cellular › Cellular Data Network.*

   → Contact your carrier for correct APN (Access Point Name).

4. **Step 4: Test with Another Device**

   → If Wi-Fi drops on other devices too → Router issue.

   → If only iPhone → proceed.

5. **Step 5: Update iOS & Carrier Settings**

   → *Settings › General › About* (wait for carrier update

prompt).

→ Restart iPhone.

Still Disconnecting?

→ Disable *Wi-Fi Assist* and Bluetooth accessories temporarily.

→ Visit an Apple Authorized Service Provider for antenna diagnostics.

App Crashes or Lag Flowchart

Problem: "Apps freeze or iPhone feels slow."

1. **Step 1: Force Close and Reopen the App**

   → Swipe up from the bottom → pause → swipe app away.

2. **Step 2: Update the App**

   → Open App Store › Profile › Update All.

3. **Step 3: Clear Cache or Reinstall App**

   → Delete and reinstall to clear corrupted data.

4. **Step 4: Restart iPhone**

   → Often fixes temporary software loops.

286

5. **Step 5: Check iOS Update**

   → *Settings › General › Software Update.*

If the Problem Persists:

→ *Settings › General › Transfer or Reset iPhone › Reset All Settings.*

→ Still lagging? Run diagnostics via *Apple Support App* or restore from backup.

Appendix B: Glossary of iPhone Terms (Plain English)

A quick, plain-English guide to the jargon Apple loves to use.

RAM (Random Access Memory):

The iPhone's "short-term memory." It temporarily stores apps and tasks you're using so they open fast. When full, your phone may slow down.

Cache:

Temporary data apps save to load faster next time. Clearing cache frees space and can fix glitches.

APN (Access Point Name):

The code that connects your phone to your mobile carrier's data network. Incorrect APN = no mobile data.

iCloud:

Apple's online storage that automatically saves photos, files, and backups.

iCloud Keychain:

Apple's password manager — it remembers login details securely across your devices.

Face ID:

Your phone's face-recognition system. It uses infrared mapping for security, not ordinary photos.

Handoff:

A feature that lets you start an email, note, or webpage on one Apple device and finish it on another instantly.

Universal Clipboard:

Copy text or an image on your iPhone, and paste it on your Mac or iPad — no cables needed.

Thermal Throttling:

A protective slowdown. When your phone gets too hot, it automatically reduces performance to prevent overheating.

Low Power Mode:

A setting that pauses background activity (like mail syncing and animations) to save battery.

AirDrop:

A wireless feature for sending photos and files to nearby Apple devices instantly.

Find My iPhone:

A locator feature to track, lock, or erase your phone if lost — even when powered off.

AppleCare+:

An extended warranty plan from Apple that covers accidental damage and battery service for a small fee.

Recovery Mode:

A system state that lets you restore your iPhone from a computer if it's frozen or stuck on startup.

Firmware:

The deep-level software that makes your hardware function. Updating iOS updates the firmware too.

Appendix C: Regional Notes

(Africa / Nigeria – Localized Advice)

Your iPhone 17 Pro is global by design, but regional networks, accessories, and repair standards can differ. This section covers the most relevant Africa/Nigeria-specific considerations for users and travelers.

1. Network and SIM Considerations

- **Dual SIM and eSIM:** All Nigerian carriers (MTN, Glo, Airtel, 9mobile) now support eSIM. For reliability, activate your eSIM at an official service center rather than through online QR codes.

- **Data Speeds:** 4G LTE is stable in urban areas, but expect occasional drops in rural zones. Use *Settings › Cellular › Voice & Data › LTE Only* for better stability if 5G coverage is inconsistent.

- **APN Settings:** Some carriers require manual entry after SIM activation. Contact your provider or visit *Settings › Cellular › Cellular Data Network* to input the APN values manually.

2. Charging and Power Supply

- **Voltage:** Nigeria uses *230V / 50Hz*, fully compatible with Apple's 100–240V adapters.

- **Surge Protection:** Always plug your charger into a surge protector or voltage regulator. Frequent power fluctuations can damage iPhone batteries over time.

- **Inverter/Power Bank Charging:** Use certified PD (Power Delivery) power banks or inverters with pure sine wave output for safe charging.

3. Authorized Repair Centers and Warranty

Apple doesn't yet have a full retail presence in Nigeria, but *authorized service providers* exist in major cities.

- **iStore Nigeria** (Lagos, Abuja): Offers warranty and AppleCare+ repairs.

- **Slot Systems & Finet**: Official partners for out-of-warranty repairs and diagnostics.

Check Apple's coverage tool here:
https://checkcoverage.apple.com

Tip: Keep your proof of purchase (digital or paper). It validates your warranty even if your device was bought abroad.

4. Regional App and Service Availability

- **Apple Pay:** Still rolling out in some African regions — may not work with all banks.

- **Siri:** Works fully in English but may not support localized accents perfectly yet.

- **Apple Music & iCloud+**: Fully available with Nigerian billing. Consider using USD or GBP billing methods for app subscriptions to avoid local conversion issues.

5. Practical Usage Advice for African Environments

- **Heat Management:** Avoid leaving your iPhone in direct sunlight or car dashboards. Ambient heat above 35°C can trigger thermal throttling and slow performance.

- **Dust Protection:** Use a high-quality case with port covers if living in dusty regions.

- **Connectivity Tips:** In poor-signal areas, enable *Wi-Fi Calling* (if carrier supports it) or use *Offline Maps* for navigation.

Final Note for Regional Users

Your iPhone 17 Pro is built to adapt anywhere in the world. Whether you're using it in Lagos, Accra, Nairobi, or Johannesburg, your experience depends less on region and more on optimization. Follow these local tweaks, register your device under AppleCare+ where available, and always back up your data before updates or service visits.

Acknowledgments

This book could not have been written without the support, patience, and insight of many people.

- **First and foremost, my readers.** Your questions, feedback, and reviews of earlier guides inspired me to create a troubleshooting book that goes beyond the basics. You are the reason this project exists.

- **My family.** Thank you for understanding the long hours I spent testing fixes, researching updates, and rewriting chapters each time Apple changed something overnight. Your encouragement kept me grounded.

- **The tech community.** Countless Apple users, bloggers, and forum contributors generously shared their experiences, problems, and solutions in online discussions. This book builds on that collective wisdom, organized into one accessible resource.

- **Independent authors and educators.** I am inspired by fellow writers who work outside of corporate publishing but still manage to produce guides that make technology more human, more understandable, and more useful.

- **Apple engineers and designers.** While this is an independent guide, I acknowledge the people at Apple who continue to push technology forward. Without their innovations, there would be no iPhone—and no need for books like this.

Finally, I want to thank you, the reader, for picking up this book. By choosing an independent guide, you're supporting authors who are passionate about clarity, practicality, and everyday usability. I hope this book earns a place on your shelf—or in your iPhone's library—as a resource you return to whenever you need answers.

www.ingramcontent.com/pod-product-compliance
Lightning Source LLC
Chambersburg PA
CBHW031841200326
41597CB00012B/227